Philip W. Robson

Power Gas Producers

Their Design and Applications

Elibron Classics
www.elibron.com

Elibron Classics series.

ISBN 0-543-69311-2 (paperback)
ISBN 0-543-69313-9 (hardcover)

This Elibron Classics Replica Edition is an unabridged facsimile of the edition published in 1908 by Edward Arnold, London.

WALTER G. PERKINS.

POWER GAS PRODUCERS

THEIR DESIGN AND APPLICATION

BY

PHILIP W. ROBSON,
OF THE NATIONAL GAS ENGINE CO., LTD.
SOMETIME VICE-PRINCIPAL OF THE MUNICIPAL SCHOOL
OF TECHNOLOGY, MANCHESTER

WITH 105 ILLUSTRATIONS

LONDON
EDWARD ARNOLD
41 & 43, MADDOX STREET, BOND STREET, W.
1908

PREFACE.

THE gas engine coupled with its own gas producer has now taken so important a position as a means of deriving cheap power that it has become necessary for the engineer to acquire a knowledge of their principles as one of those subjects with which he must essentially be familiar. It is hoped that this book may prove an aid to those who are seeking up-to-date information, not only as it provides a fairly complete memorandum of the practice at present obtaining amongst the leading English makers of power gas producers, but also in the attempt which has been made to trace that connection between theoretical considerations and actual practice which should underlie all sound design. The scheme adopted therefore in each section of the book is firstly to indicate the theory of the subject as at present understood and then to shew how that theory is applied more or less by different makers in the construction of the gas plants they manufacture, after which typical examples of actual installations are described as shewing how power gas is being introduced at the present time.

It is perhaps difficult for anyone like myself, daily engaged in furthering the interests of one firm, whose productions are constantly being offered to the public in competition with those of other makers, to impartially review the work of the latter, but I hope it will be considered that I have not greatly erred in this respect. Many firms, whose names it gives me great pleasure to mention hereafter, have most generously supplied me with much detailed information, and I cannot adequately express my sense of obligation to them for the very great courtesy they have thus shewn. Nor can I properly thank the Directors of The National Gas Engine Co. Ltd., without whose cordial approval it would have been impossible for me to have published much of the special matter contained herein, and I also take this opportunity of expressing my appreciation of the assistance rendered by Messrs Clifford Digby and W. W. Adam, the former in preparing the numerous drawings and diagrams and the latter for several useful translations from French and German papers.

I have endeavoured as an engineer to write this book for engineers, and thus prominence has been given to the various details of construction and working upon which the successful application of gas producers has been found to depend so largely, as well as to theoretical considerations. For the same reason I have adhered to the usual trade terms in expressing quantities of heat, calorific values and so forth rather than to those which are more particularly associated with purely scientific treatises.

I trust that in spite of the many omissions which limits of space unfortunately necessitate, the information given in the following pages may help to supply a want which many power users, consulting engineers, gas producer designers, factory and insurance inspectors and others have felt to exist in the past.

P. W. ROBSON.

ASHTON-UNDER-LYNE,
March 1908.

The Author gratefully thanks the following firms for kindly supplying drawings and other information in respect of their special productions:

Messrs THE NATIONAL GAS ENGINE Co. Ltd., Ashton-under-Lyne.

„ TANGYES Ltd., Birmingham.

„ THE HORSEHAY Co. Ltd., Horsehay, Shropshire.

„ THE MASON GAS POWER Co. Ltd., Manchester.

„ THE CAMPBELL GAS ENGINE Co. Ltd., Halifax.

„ W. J. JENKINS & Co. Ltd., Retford.

„ MATHER & PLATT Ltd., Manchester.

„ THE DELLWIK-FLEISCHER WATER GAS SYNDICATE, London.

and to

Prof. W. E. DALBY, M.A., B.Sc., M.Inst.C.E.

CONTENTS.

SECTION I.

SUCTION GAS PRODUCERS.

SECTION II.

PRESSURE PLANTS FOR NON-BITUMINOUS FUEL.

SECTION III.

PRESSURE PRODUCERS FOR BITUMINOUS COAL.

POWER GAS PRODUCERS.

INTRODUCTION.

During the last five years the application of gas power has received an enormous impulse. It is interesting to note that the effect of the stimulus in the manufacture of gas engines and producers thereby arising has extended equally to small units of power, say from 15 to 100 B.H.P., as well as to large gas engines of 500 B.H.P. and upwards, and that this development is everywhere in evidence in the great industrial centres of Europe. The general fact of the continued demand for small gas power units is all the more striking when it is remembered that electrical energy for power purposes is now available in most towns in this country at comparatively favourable rates; in fact it has been suggested that it is being sold at considerably less than the net cost of production in many places in order to provide a good "day load factor" for the plants installed at the Central Generating Stations. It is likewise to be remembered that the advocates of public electricity supply schemes freely prophesied, a few years ago, that every power user within their respective areas of supply would soon gladly embrace the advantages they offered, and that every other known method of deriving motive power was bound to fall into disuse.

Attention is drawn to these facts because, though great advance has undoubtedly been made in gas engine design and construction, gas power owes its advantageous position rather to the simple, effective and greatly improved gas producers which have lately been so successfully introduced than to the adoption of any new principle in engine practice, and it is no exaggeration to say that the advent of the Suction Gas Producer during the last three years has changed the whole aspect of cheap power production. For units of 15 to 300 B.H.P. these producers can be worked in this country at a fuel cost of from $\frac{1}{30}d.$ to $\frac{1}{10}d.$ per B.H.P. per hour, according to the class of fuel used: the combination of plant and engine is simple to work, to clean and overhaul, and is usually placed under the care of an ordinary attendant who frequently has also other duties to partially engage his attention.

Public competitive trials, such as those held by the Royal Agricultural Society of England at their Derby meeting in June 1906, and by the Highland and West of Scotland Society at Glasgow in 1905, to which full references are hereafter made, have likewise been organised for the purpose of demonstrating the suitability of the new gas power for general purposes, and the performance of the various plants exhibited on these occasions has been such as to call forth the fullest approbation from the judges appointed to conduct the tests.

The conclusions which may be drawn from these general facts are full of significance for the future and are bound to have an important bearing on industrial economics. In this country, at any rate, cheap power is an important desideratum. The power demand of eighty per cent. of British manufacturers is under 300 B.H.P., and steam driven units of this size are comparatively uneconomical. The choice for motive power, therefore, usually lies between electrical energy taken from the Public Supply, and a private source of power from gas producers and gas engines. The former is, broadly speaking, only available at cheap rates in the large towns, whereas the present tendency is for manufacturers to remove their works from crowded districts to open spaces a few miles away from the town areas where land is cheap and taxes low. In the latter cases gas power is specially suitable, and even in the towns there are few instances where electricity will be found to be anything like so cheap if a comparison is made on an equal basis between the two alternatives.

The position from the gas power point of view is not at present so simple when we pass to the consideration of larger units, say from 500 B.H.P. upwards. The general economy of the steam engine greatly improves, within limits, as the size of the unit increases. In gas engines there is no appreciable gain in thermal efficiency after 100 B.H.P. Hence there is not such a striking difference between the economy in fuel of large gas engines as against large steam engines as exists in the case of the smaller units just referred to. Still the difference is considerable, and where gas engines are running in the same power house as a modern steam plant consisting of large units, it has been found that the former give out just twice as much power for a given weight of fuel as can be obtained from the latter. This result is most encouraging when it is remembered that large gas engines and gas producers are practically in their infancy. Though they have been made to work with much success, everyone acquainted with the subject must realise that there are bound to be great improvements effected in their design in the future. Bearing this in mind, together with the natural tendency which exists for the price of fuel to rise, it is reasonable to suggest that the use of large gas producers and engines will be enormously extended during the next twenty years. Moreover for furnace work in dealing with heavy armour plating, general annealing and so forth, producer gas is now being largely used and increasingly employed, as it affords a means of obtaining greater uniformity of temperature, better regulation, and less oxidisation than can be attained by direct firing with solid fuel.

The efficiency of both Suction and Pressure type gas producers is high, being from 85 °/ₒ to 90 °/ₒ in the case of the former and 80 °/ₒ in the case of the latter, and, from the point of view of thermal efficiency, greater improvements must be expected in the future from the engines rather than from the producers, though there are a great many points in general practice in connection with the latter, especially in construction and durability, which require modification and improvement.

Producer gas has arrived at its present stage of commercial usefulness because its application is the embodiment of sound principles. Its use will be greatly extended in the future as those principles become better understood both in theory and in practice.

Whilst it is true that the first practical gas producer was introduced into this country from Germany by Sir William Siemens, England can claim chief credit by the work of

Mr J. E. Dowson of London, for having devised and perfected the first complete gas plant giving a cool, clean gas suitable for engine work and use with bunsen flames.

France is associated with the introduction of the Suction Gas Producer, the first of which was made by Mons. Bernier in Paris in 1894, and important pioneering work in the development of large Pressure plants for using bituminous slack has been done by Dr Mond, Mr Alfred Wilson, and others, and the different parts of the subject with which these names are respectively coupled are dealt with in the following pages.

One of the greatest difficulties experienced in introducing a new form of power arises through lack of general familiarity on the part of power users with the chief features of the apparatus which is put before them, and, as a consequence, minor circumstances frequently cause stoppages and trouble, but these, not being inherent defects, only require to be understood to be afterwards averted. In the succeeding chapters the subject will be dealt with from the point of view of those concerned with the manufacture and management of gas power plants, and the author's experience in this direction with many hundreds of producers has been embodied so far as space permits.

SECTION I.

SUCTION GAS PRODUCERS.

CHAPTER 1.

GENERAL PRINCIPLES.

It is well known that in all gas engines, whether working on the usual "Otto" or four-stroke cycle, or on the "Clerk" or two-stroke cycle, a combustible gas is drawn or pumped from the gas supply and introduced into the working cylinder of the engine. In the former case, the engine cylinder and working piston are themselves used as the pump for the time being, and on the "suction" stroke of the cycle, when the main inlet valves in connection with the gas and air supply are opened, the forward movement of the piston draws in the required charge, which is afterwards compressed on the return stroke and then ignited. In two-stroke cycle engines there is an independent pump which draws in the gas and air supply in a similar way, but thereafter discharges its contents to the working cylinder of the engine. It is to be noted that in both cases there is a suction action on the gas mains every working stroke, and if the engine is governed on the "hit and miss" principle, this action only takes place when the governor causes the gas valve to be opened, while if the governing is on the principle of throttled or graduated charges, the degree of suction during each working stroke is in strict proportion to the load. In other words, whatever be the method of governing adopted with the engine, a suction action, proportionate in its degree to the load the engine is called upon to meet, is exercised on the gas main through which the gas supply is brought to the engine.

It follows, therefore, that if a direct connection be made between a gas producer and an engine so that all the gas made must pass straight to the engine without an intervening gas holder and the engine consequently sucks direct on the generator of the producer, the suction action of the engine will automatically regulate the amount of gas evolved from the producer to suit the load which the engine is called upon to meet. This arrangement is found to work well in practice, and the type of gas producer required for the purpose is called the **Suction Gas Plant.**

The general appearance of the combination of a gas engine with its suction gas plant is illustrated in Figs. 1 and 2, which

FIG. 1. National gas engine and Suction gas producer.

shew typical apparatus as made by the National and Tangye firms respectively.

The cooling water tanks for the engine cylinder jacket are shewn in connection with the Tangye combination, and it is of course to be understood that similar water cooling arrangements are required for all types of engines.

Fig. 3 shews a sectional diagram how the National producer is connected with the engine. It may be here briefly mentioned that the fuel to be used with these plants is Welsh or Scotch anthracite, charcoal, or, alternatively, a fair quality of ordinary

FIG. 2. Tangye gas engine and Suction gas producer.

gas coke, and that the air entering the furnace of the gas generator must be partially saturated with water vapour. The figure shews how, what would be otherwise, the waste sensible heat of the outcoming gas is utilised to heat the vapouriser wherein the incoming air catches up its moisture.

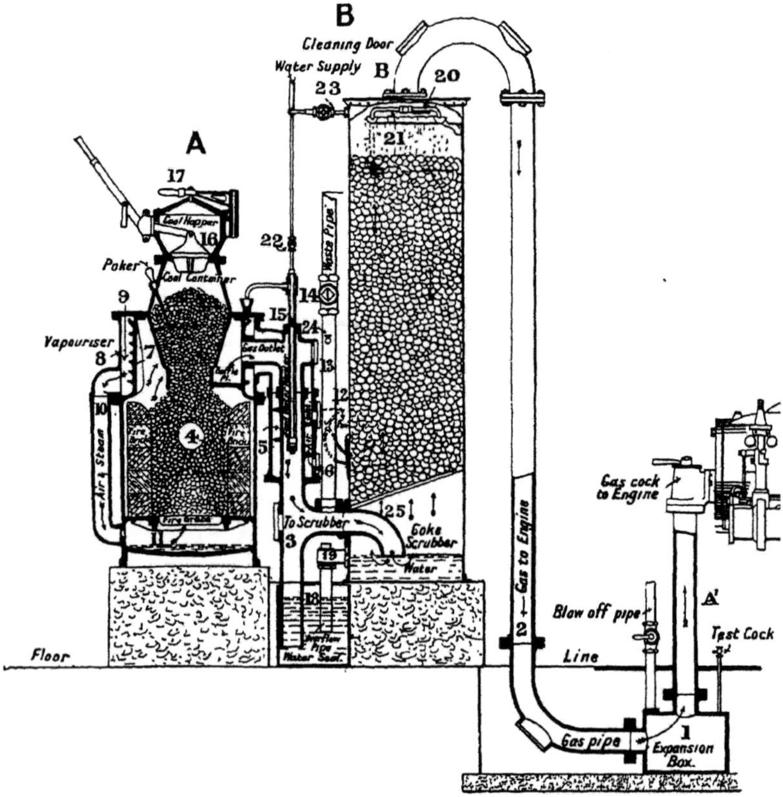

FIG. 3. Section through National Producer shewing connection to engine cylinder.

The suction gas making plant as described in Fig. 3 above consists essentially of the furnace "A," in which the fuel is burnt, and "B," which is a long cylindrical vessel filled with damp coke through which the gas is passed and consequently cooled and cleaned on its way to the engine. The furnace "A" is usually called the **generator**, and the wrought iron cylinder "B" containing the damp coke is called the **scrubber.**

When the gas making plant is at work in conjunction with the engine, the general action is as follows:

(*a*) The engine draws its charge of gas from the expansion box (1) which is directly connected with the top of the scrubber by the gas main (2). The scrubber in turn receives its supply of gas from the gas outlet pipe (3) which connects the producer to the scrubber, and this outlet pipe is connected in such a way as to draw off from the producer the gas which is made through the partial combustion of the fuel in the furnace (4) of the producer.

(*b*) It will be thus seen that every time the engine sucks in a charge of gas, the suction action is communicated from the engine through all the interior connections of the gas plant until it is felt right at the furnace of the producer. A definite air inlet is provided to allow the air required for combustion to be drawn into the furnace of the producer at each suction stroke so as to make additional gas to replace that drawn off by the engine, and consequently the production of the gas is quite automatic and in accordance with the demand made by the engine, which in turn is regulated by its governor.

(*c*) For the proper production of gas and the good working of the producer steam must be mixed with the air passing to the furnace so as to keep down the temperature of the latter, otherwise the firebars would be burnt out and the body of the generator cracked.

(*d*) As the gas which comes off at the upper part of the producer is at a considerable temperature, it is used to vapourise the water required for the steam supply. In the plant shewn in Fig. 3, therefore, the air and steam supply is arranged as follows:

A jacket (5) is provided round the gas outlet pipe and, under the suction effect of the engine already referred to, air passes in at the inlet (6) and, gradually circulating round the gas outlet pipe, is heated considerably before it passes into the vapouriser. The vapouriser is formed by the internal circular shell (7) and the external circular shell (8), an annular space existing between them. The inner shell (7) is heated by the outgoing hot gases coming in contact with its interior on which heat-catching ribs are cast. On its external surface a supply of heated water is continually fed and is evaporated by the heat of the surface. There is consequently an annulus (9) which is always kept full of steam while the plant is at work.

(*e*) As soon, therefore, as the entering air, which has already been heated by its passage through the air jacket (5), reaches the vapouriser, it becomes saturated with steam in passing round the vapouriser on its way to the air and steam pipe (10). This latter accordingly feeds the space underneath the grate with a mixture of air and steam, which duly passes through the fire.

(*f*) There are a few additional important details which are required for working the plant, namely, the fan (12) which is used for blowing in air when starting the plant, *i.e.*, before the engine is got to work. When the fan is being used, the blow off pipe (13) is brought into action by opening the cock (14). This pipe is extended to the outside air, and the cock (14) is shut as soon as the engine is got to work. (15) is a simple water heating arrangement which takes further advantage of the waste heat in the outcoming gas from the producer. Double valves (16) and (17) are necessary in the coal hopper, through which the coal is introduced to the inside of the producer. Valve (16) is kept closed while the lid valve (17) is open and the fuel poured into the coal hopper. The lid valve (17) is then replaced and the hopper valve (16) is opened, the fuel consequently dropping through. It is essential that no air should enter the producer when at work, excepting in the appointed way through the air supply pipe (10) and from thence through the fire grate.

(*g*) In connection with the scrubber there is a seal pot (18) into which the overflow pipe (19) discharges the waste water from the scrubber which is continually used whilst the plant is at work for cleaning and cooling the gas. This water is fed into the scrubber by the sprinkler pipe (20) and it is spread over the whole surface of the coke by the distributing dish (21).

Prof. Dalby, in a paper which he read before Section G of the British Association at their York Meeting in August, 1906, summarised the general action of the Suction Producer in a very clear and admirable manner, and the following extract from this paper will further explain the method of working the apparatus:

"Fig. 4 is a diagrammatic representation of a plant. The gas current through the plant is put on the rack as it were, and stretched out to a straight line. To the right of the diagram will be seen the engine cylinder, which acts as a pulse to the system. When the piston moves out to the right the pressure is reduced all through the plant, and air is drawn in at the points A and B. Confining our attention to the air entering at A, it makes its way through what may be called the 'tube' and receives heat in the region marked H, thus increasing its capacity to absorb moisture. Water enters in this region,

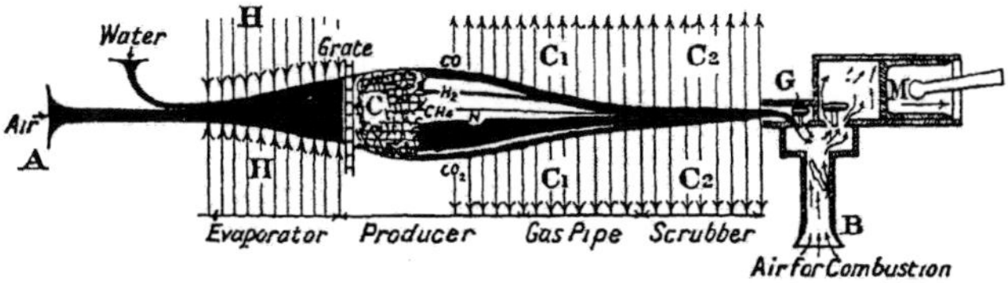

FIG. 4. Diagram of Suction Plant action.

and the heated air, acting as a sponge, saturates itself with water vapour, and then passes on to the grate of the producer, where it passes through an anthracite or coke fire. It will be observed that at this stage the quantities entering into this process for 1 lb. of carbon are 4·5 lbs. of air and about 0·084 lbs. of water. Chemical changes now take place and the air and the absorbed moisture become transformed, after the passage through the fire, into a gas containing roughly 29 lbs. of carbonic oxide, 1½ lbs. of hydrogen, ½ lb. of methane, 57 lbs. of nitrogen, and 12 lbs. of carbon dioxide per 100 lbs. The density of the gas in this state is small, and therefore, if it were taken direct into the cylinder, a charge of small energy value would be obtained. It is, therefore, necessary to cool the gas down to something like the atmospheric temperature during its passage through the region C 1. C 2. The gas then, in some cases, emerges into an expansion chamber placed close to the engine. Examining, now, more particularly the cylinder end of the apparatus, and assuming a suction stroke to be taking place, the gas enters through the valve G into a chamber which is in direct communication with the air, consequently the suction produced by the piston draws into this chamber, and from thence into the cylinder, not only gas from the producer through the valve G, but the quantity of air necessary to make it into an explosive mixture through the air pipe B. The quantity of air through the air pipe B is controlled by a valve which is set in a definite position. The mixture of gas and air passes into the cylinder through the valve M, and is compressed during the return stroke of the engine until, at a suitable point, it is fired by an electric spark obtained by breaking the circuit through which a current flows, the current being derived from a magneto of the Simms Bosch type, mounted on the side of the engine, and operated by a pin on the cam shaft. Thus the movement of the piston to the right causes a flow of air and steam into the apparatus at the left, which undergoes a series of complicated chemical reactions in the producer part, and at the same time draws gas from the apparatus and combines it with a suitable mixture of air to form the explosive mixture. At the end of this suction stroke the producer is cut off from connection with the engine by the closing of the valves G, M, and the remaining three strokes of the Otto cycle are completed. These operations are all perfectly straightforward, and present no apparent difficulties. But the extraordinary feature of the suction plant is that an engine can go on working continuously, and these operations can go on minute after minute, hour after hour, and day after day, so that the explosive mixture made in the mixing chamber, and the power developed by the cycle of operations in the cylinder, are automatically regulated to enable the engine to run on a variable load without the necessity of altering the position of a single valve or handle.

It will be observed (Fig. 4) that heat is taken into the system in the region H, and is expelled again in the regions marked C 1 and C 2; and it will be at once apparent that if heat is to be taken away from region C, and is to be introduced in the region H, the apparatus may be so constructed that the heat required at H is that taken from the heat rejected at C, thus introducing the principle of regeneration."

The chemical action and general theory will be dealt with later, but meanwhile it will be noted that the gas evolved from such a producer contains, roughly, 40 to 45 °/₀ of combustible gases together with 60 to 55 °/₀ of diluents, the chief proportion of which is the nitrogen contained in the air passing through the furnace of the producer, and which is not reacted on. In practice this gas evolved from such a producer has a calorific value of from 130 to 145 B.Th.U.'s per cubic foot, and is capable, in a well designed engine, of giving effective mean pressures of from 80 to 85 lbs. per sq. inch, when using anthracite as fuel. With a particularly good class of fuel and a producer in good order, the author has frequently been able to maintain mean pressures in the engine of over 90 lbs. per sq. in. with compression pressures of 120 lbs., but this is exceptional, and in computing the ordinary working load of the engine it will not be found advisable to count on mean pressures of over 80 lbs. per sq. in.

General theoretical considerations. The function of any power gas producer is to change the character of solid fuel so that the maximum volume of combustible gas of as high a calorific value as possible may result therefrom. To carry this into effect, the necessary degree of temperature is obtained in the producer to gasify the fuel by a process of partial combustion of the latter, but, obviously, as little as possible of the heating value of the constituents of the fuel should be absorbed in the producer. The primary chemical reaction which takes place is due to the combination of the oxygen of the atmosphere with the carbon of the fuel to form carbon monoxide in accordance with the following equation:

$$2C + O_2 = 2CO \quad \text{.......................}(a).$$

It can be shewn that the heat which the carbon monoxide so formed represents is only 70 °/₀ of the total heat of the carbon acted on, so that the remaining 30 °/₀ is liberated in the producer. The physical effect of this, if allowed to be cumulative, would be to produce a very high temperature in the producer, if designed on the usual proportions, and the gas passing off would consequently be exceedingly hot. As a matter of fact, the working temperature of the producer under such conditions would be considerably over 1500° C. At such a temperature any portion of ash there might be in the fuel would be fused into clinker, thus blocking up the fire and vitiating the action of the producer, and, furthermore, the loss of heat would obviously be very considerable if the gas had to be cooled for engine work before being used.

We thus see the serious limitations there are in working a producer with air only, and though Siemens made his first producers on this principle and worked them with comparative success, it was principally due to the fact that he could use the gas made whilst it was still hot, and it was not necessary for him to cool it down as is required for engine work.

Having regard to the fact that air contains $\frac{4}{5}$ of nitrogen by volume and that this nitrogen is not acted upon as it passes through the producer, it follows that all producer gas contains a large proportion of nitrogen which averages down the calorific value per unit volume of the gas made. Under the conditions named, the actual calorific value of the gas produced would be about 110 B.Th.U.'s per cubic foot, and this is scarcely high enough for use with gas engines working with a usual ratio of compression.

At this point, therefore, three important practical requirements are presented which may be stated as follows:

(*a*) It is necessary to reduce the temperature of the fire to prevent the formation of clinker.

(*b*) The gas made must be enriched so as to make it suitable for use in an engine.

(*c*) Some effort must be made to turn to useful effect as much as possible of the 30 °/ₒ of the heat of the fuel which is liberated in the producer.

Mr Dowson was the first to shew how these points could be successfully met by introducing a steam blast along with the air. Steam in conjunction with hot carbon acts in either of the following ways:

$$H_2O + C = H_2 + CO \quad \ldots\ldots\ldots\ldots (b),$$

$$2H_2O + C = 2H_2 + CO_2 \quad \ldots\ldots\ldots\ldots (c).$$

In a gas producer, possibly both actions take place to some extent, and it is to be specially noted that both require heat to enable them to take place, *i.e.* the heat represented by the hydrogen and carbon monoxide gases on the right hand side of equation (*b*) is greater than that possessed by the carbon on the left hand side, and the balance is supplied by the heat liberated in accordance with equation (*a*). Similarly in equation (*c*) the heat represented by the hydrogen evolved is greater than that of the carbon, though in a less degree than in reaction (*b*).

Hence we may say briefly that the effect of introducing a constant jet of steam into the producer is to absorb the heat which is being continually liberated in the production of carbon monoxide by the principal reaction represented by equation (*a*).

This continual absorption of heat reduces the general temperature of the producer and goes a long way towards meeting the first of the three difficulties stated above which were found with the making of air gas.

We next have to consider the effect of the reactions (*b*) and (*c*) on the heating value of the gas produced when the steam jet is employed. In the first place it is to be borne in mind that *the total potential heat available is that in the fuel only*, and clearly this is not affected whether steam is passed into the producer or not. Still the gas produced *is* enriched by using steam, for the hydrogen and carbon monoxide formed by the reactions described are produced independently of the air supply, and hence, after their formation, they displace a proportionately larger volume of the inert nitrogen which is present in the gas produced by reaction (*a*). The effect of the steam is, therefore, cumulative, for it reduces the temperature of the fire and prevents the formation of clinker: it forms free hydrogen and carbon monoxide, the presence of which in the gas produced causes a smaller percentage of nitrogen, and the gas has, therefore, a higher heating value per unit volume. The heat absorbed by the splitting up of the steam is the heat liberated in the primary reaction for the production of carbon monoxide, and would otherwise be lost.

Another useful effect of this general reduction of temperature is that the gas leaves the producer at a lower temperature and consequently requires less cooling water to be passed through the scrubbers than would otherwise be the case.

In the Suction type of plant at present under consideration the sensible heat in the outcoming gas is utilised to vapourise the water necessary for the steam supply, so that the use of a separate boiler detached from the generator is obviated. It is found in practice, that in a well designed plant, quite sufficient heat may be obtained in this way to produce all the steam required, and since it has been shewn that the effect of this steam, which is generated gratuitously so to speak, is beneficial to the production of gas in every way, it follows that the efficiency of the producer is considerably higher than if the steam be not used. We have already indicated that the maximum efficiency in the latter case is 70 %, but with the use of steam generated on the principle described above the efficiency is raised to over 85 %.

The actual heat quantities involved in the various reactions which take place in the producer and upon which the proper action of the latter depends, are fully dealt with subsequently in Chapter 2, and should be studied most carefully; but the general considerations indicated in the foregoing pages will be found sufficient to illustrate the chief points which have to be borne in mind in designing and working suction producers.

Before proceeding to discuss the actual design of such producers, there are several important considerations in respect of the character of the fuels which alone are suitable for use with them which must be clearly understood, for, naturally, the character of the fuel to be dealt with directly influences the design of the generator and scrubbers of the plant. The following statements must, therefore, be carefully followed:

Fuel requirements for ordinary suction producers. Up to the present time suction producers are only employed in this country on a commercial scale for use with anthracite coal or coke, and they are principally applied for power units of from 10 B.H.P. to 150 B.H.P. They have been successfully made up to 300 B.H.P., however, and the author has designed several installations of this latter size which are working well and giving every satisfaction to the user. The reason why the fuels named are alone suitable for the purpose of suction producers will be understood by comparing the average composition of the several different classes of fuel usually available as follows:

TABLE 1.

Approximate Compositions of Different Fuels.

	Carbon	Volatile matter	Ash
Anthracite	92 %	6	1·5
Non-caking bituminous coal	70 %	20	8
Gas coke	85 %	6	9

From the foregoing table it will be seen that anthracite contains a high percentage of carbon and a small percentage of volatile matter and ash: gas coke has considerably less carbon, approximately the same amount of volatile matter, and a larger percentage of ash: bituminous coal may have only two-thirds the amount of carbon as compared with anthracite, while its percentage of volatile matter is very high, and the amount of ash may likewise be most troublesome.

The two factors, percentage of volatile matter and of ash, have the most influence in the successful working of gas producers, for in the case of bituminous coal and gas coke the volatile substances in the coal usually distil and are carried off from the producer in the form of tarry vapours, which are most difficult to clean out of the gas, and, in the case of producers using bituminous fuel, the scrubbing arrangements require to be very elaborate indeed as compared with the simple scrubber shewn in Fig. 1 which is quite sufficient when anthracite is used in the producer. As between anthracite and gas coke, which have practically the same proportion of volatile matter, there is this difference, that that given off from the coke is of a more tarry nature, and an additional sawdust scrubber or tar extractor is usually added when the latter fuel is to be used constantly.

The amount of ash present in the fuel is a point of importance in so far as it fuses together under the influence of the high temperature of the interior of the producer and forms hard clinker, which latter may cause great trouble by blocking up the fire and preventing the necessary reactions for the production of good gas taking place.

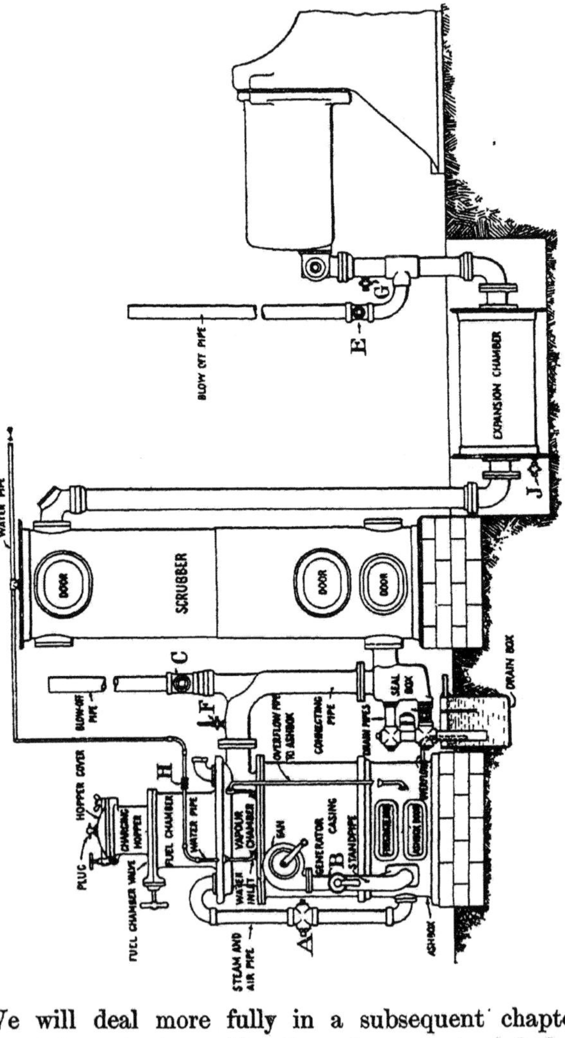

FIG. 5. Diagram of Tangye Suction Plant.

We will deal more fully in a subsequent chapter with the most important considerations in respect of fuel for gas producers, but for the present we will briefly indicate the significance of the points before given in respect of the suction producer:

1. If the volatile matter appears as tar in the gas, as is the case when ordinary bituminous slack is used, an extensive scrubbing apparatus must be employed to extract it, as this tar

FIG. 6. 150 B.H.P. National Suction Plant.

causes the engine valves to stick, and leads to other difficulties with the engine which cannot be tolerated. On the other hand, however, if extensive scrubbers be used, too much back pressure is put on the engine during the suction stroke caused by the resistance to the passage of the gas which these scrubbers offer, and consequently the engine is prevented from drawing in the proper weight of charge and it fails to work in a desirable way.

2. If the ash forms a vitreous clinker, trouble will be caused by the latter becoming attached to the sides of the interior of the producer and gradually growing therefrom until the effective dimensions of the furnace of the producer are so restricted as to interfere with the production of a sufficient quantity of good gas.

3. It follows from these considerations, taken in conjunction with the average fuel compositions given in Table 1 on page 17, that *anthracite coal and good clean coke* or charcoal are the only suitable fuels for use in a suction producer of the general type shewn in Fig. 1.

Fig. 5 shews an outside elevation of the Tangye producer illustrated in Fig. 2, and the usual technical names by which the various parts are designated are respectively indicated in connection therewith. C and E are the blow-off cocks used in starting up the plant (see Chapter 8), F and G are test cocks for trying the quality of the gas, cocks A and B are respectively for shutting off the vapouriser when blowing with the hand fan and for closing off the fan when working by suction. H is a small valve for regulating the water feed to the vapouriser.

Fig. 6 illustrates the latest form of suction plant suitable for serving a 150 B.H.P. engine as made by the National Company, but the same general principles which we have described in this chapter are applied to all sizes of producers.

CHAPTER 2.

REACTIONS IN THE PRODUCER.

It is essential to a right understanding of the principles of design, working and management of gas producers that the chief reactions which take place in the producer should be appreciated. Before investigating these reactions, and the results flowing from them, it is necessary to recall a few simple chemical terms and values such as may be obtained from any standard work on elementary chemistry.

Heat units. The British Thermal Unit (written B.Th.U.) is the quantity of heat required to raise the temperature of one pound of water through one degree Fahrenheit.

Calorific power or value. The calorific power of a solid is the total *quantity of heat* evolved by the combustion of unit mass of the material starting from the atmospheric temperature and cooling back to the same. In the case of a gas, unit volume is taken under the same conditions in determining the calorific value. In this country, the heating or calorific value of a gas is usually expressed in commerce in British Thermal Units per pound for solids, and in British Thermal Units per cubic foot for gases.

Higher and lower calorific values. In referring to the heating values of power gas it has become customary to distinguish between the "higher" and "lower" heating values. The reason is that when such gases burn, the products of combustion are chiefly carbon dioxide and water with the result that when in an investigation of the calorific value these products are cooled back to atmospheric temperature, the water vapour condenses and in so doing yields up its latent heat. If, however, the gas is used in an engine, the products of combustion pass away to the exhaust at high temperature and consequently the latent heat in the water vapour referred to is not given up. The calorific values of the gas with and without this latent heat are termed the "higher" and "lower" calorific values respectively.

TABLE 2.

Selected Calorific Values.

	B.Th.U. per cubic foot		B.Th.U. per lb.
	Higher Value	Lower Value	
Carbon monoxide, CO ...	342·4		
Hydrogen, H	347·1	292·3	
Ethylene, C_2H_4	1713·0	1603·0	
Methane, CH_4	1072·0	963·0	
Carbon, C			14650

Molecular weights. The molecular weights of different gases are the respective comparative weights of the same volumes of such gases at standard temperature and pressure. Chemical formulae are used to express the composition of substances and always represent quantities of the same equal to the molecular weights in grams, kilos, or pounds. For instance, in considering the bearing of a chemical equation the respective molecular weights may be substituted for the various formulae employed, and these weights may be expressed in any of the units named above providing that the same unit is consistently employed throughout the equation.

The molecular weight of a gas in pounds occupies 357·5 cubic feet.

TABLE 3.

Selected molecular weights.

Name	Chemical Formulae	Molecular weights
Carbon dioxide	CO_2	43·9
Carbon monoxide ...	CO	27·94
Ethylene..............	C_2H_4	27·95
Methane	CH_4	15·97
Hydrogen	H_2	2·00
Oxygen	O_2	31·93
Nitrogen..............	N_2	28·02
Water vapour........	H_2O	17·96

The action of air on incandescent carbon. Air consists broadly of one-fifth oxygen and four-fifths nitrogen by volume and, as in the combustion of carbon in air the nitrogen remains unchanged, we have to consider the action of the oxygen only on the carbon, but we have also to take account of the volume of nitrogen which passes through the vessel in which combustion takes place without being changed. It has been usually accepted that carbon burns directly to carbon dioxide, which is immediately afterwards reduced to carbon monoxide through being in contact with an excess of incandescent carbon, this being a process of partial combustion of carbon by air in an enclosed vessel. These two reactions are thus expressed:

$$C + O_2 = CO_2 \quad \ldots\ldots\ldots\ldots (a),$$

$$CO_2 + C = 2CO \quad \ldots\ldots\ldots\ldots (b).$$

Chemists are, however, by no means agreed that the production of the carbon monoxide is brought about in this way, and there appear to be good grounds for thinking that under the circumstances named, the carbon really burns directly to carbon monoxide thus:

$$2C + O_2 = 2CO \quad \ldots\ldots\ldots\ldots (c).$$

In any case, it is to be noted that the final result is the same, for the combined result of equations (*a*) and (*b*) is precisely that of the single reaction expressed by equation (*c*), the relative amounts of carbon and oxygen used, and volume of monoxide produced, being identical in the two cases. In, therefore, considering the heat quantities involved in the production of carbon monoxide from carbon, it is sufficient to confine our attention to the reaction expressed by equation (*c*). The atomic weight of carbon is 12, and hence we may in this equation assume each part of carbon to be 12 lbs. From the definitions given as to the volume occupied by the molecular weight of a gas in pounds, and also from Table 2 of calorific values, we may now put down the various particulars involved in the reaction referred to as follows:

TABLE 4.

	2C +	O_2 =	2CO
Weights in lbs. involved in equation	2 × 12 = 24	31·93	2 × 27·94
Volume of gas produced. (Since the oxygen is derived from air, it carries with it four times its volume of nitrogen, which is not reacted on.)		357·5 O + 1430·0 N	2 × 357·5 = 715 c. ft. CO 1430 c. ft. N
Heat units in B.Th.U.'s .... (Carbon = 14650 B.Th.U.'s per lb.) (CO = 342·4 B.Th.U.'s per c. ft.) ...	24 × 14650 = 351600		715 × 342·4 = 244816

The composition of the gas produced will be accordingly:

CO...... 715 c. ft.	or 33 %	by volume
N ......1430 c. ft.	or 67 %	"
2145	100	

Total volume of gas produced = 2145 c. ft.

Calorific value $\frac{244816}{2145} = 114$ B.Th.U.'s.

Volume of gas per pound of carbon $= \frac{2145}{24} = 89{\cdot}5$ c. ft.

B.Th.U.'s per pound of carbon $= \frac{244816}{24} = 10200$.

From the foregoing it will be seen that the heat in the carbon monoxide produced is only 70 % of the total potential heat contained in the carbon, and hence if air only were allowed to react on the fuel in the producer, the efficiency of the system would be low, due to the liberation of 30 % of the available heat of the carbon which does not reappear in the gas made. This continual liberation of heat would also cause excessive heat in the producer, which would have a very bad effect on its good working.

The action of steam in the producer. In suction plant practice this 30 % available heat which is liberated from the primary reaction just described is partially used to generate steam which, being mixed with the air passing to the furnace, is decomposed by contact with the incandescent carbon, this being arranged as described in Chapter 1. The decomposition of the steam which may take place in accordance with either or both of the following equations requires a further portion of the 30 % of heat available.

$$C + 2H_2O = CO_2 + 2H_2 \quad \ldots\ldots\ldots\ldots\ldots\ldots (d),$$

$$C + H_2O = CO + H_2 \quad \ldots\ldots\ldots\ldots\ldots\ldots (e).$$

It may here be said that the reaction (*d*) takes place when the temperature of the fire is comparatively low, about 600° C., and at temperatures at from 900 to 1000° C. and upwards the alternative reaction (*e*) results. Between 600 and 1000° C., both actions take place to an extent, the second reaction gradually predominating as the temperature rises.

We may investigate the heat quantities involved in these two reactions on similar lines to those adopted with the primary reaction already considered, but as in suction producers the steam has to be raised from water at about 60° F., and completely vapourised by the surplus heat available, we must take account of this heat required preliminarily for steam raising purposes. It will, of course, be the sum of the sensible and latent heats from 60° F. which is equal to:

$$(212-60)+966 \text{ B.Th.U. per pound of water} = 1118 \text{ B.Th.U.'s}$$

for each pound of water used.

TABLE 5.

	C	+ $2H_2O$	= CO_2	+ $2H_2$
Weight in pounds involved in equation	12	$2\times18=36$	44	$2\times2=4$
Volume of gas produced			357·5	$2\times357{\cdot}5$ $=715$
Heat quantities in B.Th.U.	12×14650 $=175800$	$-(36\times1118)$ -40250		$715\times292{\cdot}3$ $=209000$
	135550			

It will be seen on comparing the heat quantities on the two sides of the equation in the above reaction that the heat value of the hydrogen produced is $209000-135550=73450$ B.Th.U.'s more than the heat supplied by the carbon and steam reacted on. This additional heat can only be obtained from that liberated by the primary reaction according to equation (*c*). We have already seen from the analysis of this equation that when 24 lbs. of carbon are decomposed to carbon monoxide 107100 B.Th.U.'s are liberated and are thus available to allow the further reaction between steam and carbon now under consideration to take place.

We have also seen that in the reaction according to equation (*d*) 73540 B.Th.U.'s are required to decompose each 12 pounds of carbon and 36 pounds of steam, and, therefore, to completely utilise the heat available from reaction (*c*) we must increase all the quantities in equation (*d*) in the ratio of $\frac{107100}{73450}=1{\cdot}46$.

Taking the equations (*c*) and (*d*) together on these lines we have the following combined result:

TABLE 6.

	$2C$ .... + O_2 = $2CO$ .... (c) $1{\cdot}46C + 2{\cdot}92H_2O$ .... = ... $1{\cdot}46CO_2 + 2{\cdot}92H_2$ .. (d)					
Weights in lbs. involved in equation	3·46 × 12 = 41·52	2·92 × 18 = 52·6	2 × 16 = 32	2 × 28 = 56	1·46 × 44 = 64·2	2·92 × 2 = 5·84
Volume of gas produced			357·5 O + 1430·0 N	357·5 × 2 = 715	357·5 × 1·46 = 522	357·5 × 2·92 = 1043
					+ 1430 N	
Heat quantities in B.Th.U.	(14650 × 41·52) = 608000	−(1118 × 52·6) − 58000		(342·4 × 715) = 245000		+(292·3 × 1043) + 305000
	550000				550000	

The composition of the gas produced will be:

CO	715 c. ft.	or	19·25 % by volume
CO_2	522 "		14·10 % "
H	1043 "		28·15 % "
N	1430 "		38·5 % "
	3710		100·00

Total volume of gas produced = 3710 c. ft.

Calorific value $= \frac{550000}{3710}$ = 150 B.Th.U.'s per c. ft.

Volume of gas per pound of carbon $= \frac{3710}{41{\cdot}52} = 89{\cdot}5$ c. ft.

Weight of steam per pound of carbon $= \frac{52{\cdot}6}{41{\cdot}52} = 1{\cdot}265$ lbs.

B.Th.U.'s per pound of carbon $= \frac{550000}{41{\cdot}52} = 13200.$

Air required per pound of carbon $= \frac{1788}{41{\cdot}52} = 43$ c. ft.

The foregoing results give a perfect heat balance, the whole heating value of the carbon acted upon being accounted for, and a similar investigation will now be made into the circumstances which arise when the steam reacts in accordance with equation (*e*).

TABLE 7.

	C +	H_2O =	CO +	H_2 ... (*e*)
Weight, in lbs., involved in equation	12	18	28	2
Volume of gas produced			357·5	357·5
Heat quantities in B.Th.U.'s	(12 × 14650) = 175800	− (1118 × 18) − 20100	(342·4 × 357·5) = 122400	+ (292·3 × 357·5) + 104500
	155700		227000	

From the foregoing figures it will be seen that the heat represented by the calorific value of the carbon monoxide and hydrogen formed is greater than the heat supplied by (227000 − 155700) = 71300 B.Th.U.'s. The latter is obtained, as in the previous case, from the heat liberated from the primary reaction. This, as we have seen, is 107100 B.Th.U.'s per 24 lbs. of carbon acted upon, so that on this basis the quantities represented by equation (*e*) may be increased in the proportion $\frac{107100}{71300} = 1·5$. For a complete heat balance we may therefore combine equation (*c*) and (*e*) as follows:

TABLE 8.

	$2C$ .... $1·5C$	+ + $1·5H_2O$	O_2 = =	$2CO$ ..... $1·5CO$ +	.... $1·5H_2$	....(*c*) ....(*e*)
Weights involved in lbs.	3·5 × 12 = 42	1·5 × 18 = 27	32	3·5 × 28 = 98	1·5 × 2 = 3	
Volumes of gas produced			357·50 + 1430·0N	357·5 × 3·5 = 1250	357·5 × 1·5 = 536	+ 1430N
Heat quantities in B.Th.U.'s	(14650 × 42) = 615000	− (1118 × 27) − 30000		(342·4 × 1250) = 428000	+ (292·3 × 536) + 157000	
	585000			585000		

The composition of the gas produced will be:

CO	......1250 c. ft.	or 39 °/₀	by volume
H	...... 536 „	16·7 °/₀	„
N	......1430 „	44·3 °/₀	„
	3216	100·0	

Total volume of gas produced = 3216 c. ft.
Calorific value = $\frac{585000}{3216}$ = 182 B.Th.U.'s per c. ft.
Volume of gas per lb. of carbon = $\frac{3216}{42}$ = 76·5 c. ft.
Weight of steam per lb. of carbon = $\frac{27}{42}$ = 0·64 lbs.
B.Th.U.'s per lb. of carbon = $\frac{585000}{42}$ = 13900.
Air required per lb. of carbon = $\frac{1788}{42}$ = 42·6 c. ft.

In both these hypothetical cases just considered the whole of the potential heat of the carbon appears in the heating value of the gas produced, *with the exception* of that required for steam raising purposes, and the only difference in their relative efficiency arises from the fact that in the first case more steam is used per pound of carbon and consequently more heat is required for raising this steam. Otherwise, though the calorific value of the gas produced according to the first combination of reactions is less than with the second combination, there is a greater yield of gas per pound of carbon in the former case, so that the B.Th.U.'s in the gas produced per pound of carbon are only 5 °/₀ less. *Up to this point*, therefore, we may say that there is very little difference in the efficiency of the two alternative ways in which the steam may react on incandescent carbon. As a matter of fact, however, other important considerations arise which greatly increase the efficiency of the second combination of reactions and which must now be taken into account.

Reversible reactions influencing general efficiency. *The action of steam on carbon monoxide.* At temperatures over 500° C. steam may react with carbon monoxide and hydrogen with carbon dioxide, these reactions being reversible thus:

$$CO + H_2O \rightleftarrows CO_2 + H_2.$$

The reduction of carbon monoxide to dioxide by this reaction is naturally accompanied by evolution of heat, and though in the scope of this work the exact conditions under which this reversible action takes place cannot be adequately discussed it may be stated that a temperature of 900—1000° C. and upwards favours the formation of carbon monoxide rather than dioxide and hydrogen. Hence for the richest gas and highest efficiency the temperature should be kept as high as is consistent with the good working of the producer in other respects, such as freedom from formation of clinker, etc.

Reactions between carbon monoxide and carbon. There is a further reversible reaction which is likewise dependent on temperature and which may be expressed thus:

$$C + CO_2 \rightleftarrows 2CO.$$

The physical meaning of this equation is that when proportions of carbon monoxide and dioxide are present together with incandescent carbon there is a tendency for the monoxide to form dioxide and *vice versa*. Naturally the destruction of the monoxide after its formation should be avoided as far as possible, and it has been proved that the tendency to change to dioxide is a consequence of comparatively low temperatures and that with temperatures of from 900 to 1000° C. the monoxide is practically in equilibrium. As the temperature falls, the reverse action producing carbon dioxide is favoured.

Conclusions. From the foregoing remarks it will be seen that the temperature at which the producer is worked has a very important influence on the quality of the gas made and also on the efficiency of the system. The working temperature is, within practical limits, entirely controlled by the amount of steam passed through the fire, and we are now able to state the conditions which should be aimed at to secure the best results:

1. Of the two alternative reactions of steam on carbon

$$C + 2H_2O = CO_2 + 2H_2,$$

and

$$C + H_2O = CO + H_2,$$

the latter yields the better gas, is more efficient, and uses less steam. It consequently allows a higher temperature in the producer, and in fact this reaction can only take place at a temperature of about 1000° C. This second reaction, therefore, is to be preferred above the first because:—

2. The temperature of this reaction also favours the maintenance of the carbon monoxide in equilibrium and prevents

the reversible actions producing carbon dioxide, which have just been described, from taking place.

3. The steam supply to the producer should, therefore, be kept well under control so that the working temperature can be kept as high as possible consistent with the prevention of the formation of excessive clinker in the furnace.

4. An excess of steam lowers the temperature all round and is generally against efficiency and prevents the production of a good gas.

5. In comparing the foregoing theoretical results with practice, it must be borne in mind that they are only approximations to working conditions, and that their value is to shew the general direction in which to aim in producer design. The importance of designing the vapourisers of suction producers to effectively control the steam supply to the fire so as to obtain the best results is so imperfectly understood that of the numerous producers which have been placed on the market there are probably not more than two in which the required control is arranged to be at all possible, and for this reason alone the conclusions arrived at should be fully appreciated at this stage. It is further to be remembered that ordinary anthracite or gas coke is not pure carbon, and that the whole of the sensible heat produced in decomposing the fuel to carbon monoxide cannot be recovered as has been assumed in our investigations. The gases leave the producer at a temperature of about 600° F., and the additional apparatus required to abstract further heat from them would be too costly and cumbersome to warrant their adoption.

6. As a pound of fuel for a suction producer will contain not more than about 90 °/ₒ of carbon, and having regard to the incomplete recovery of the waste heat, it may in any case be taken that not more than 0·8 pounds of water per pound of fuel is required to give the best results. The amount of air required for the reaction, as we have seen, is about 42 c. ft. per pound of carbon. The weight of air is 12·35 c. ft. per pound, so that each $\frac{42}{12{\cdot}35} = 3{\cdot}4$ pounds of air passing to the furnace should carry 0·80 pounds of water in suspension, or $\frac{0{\cdot}80}{3{\cdot}4} = 0{\cdot}24$ pounds

of water per pound of air. This is equivalent to saturation at about 85° F. In a well designed plant the temperature of the air passing to the furnace will be as high as 160° F. through the air being preheated in the vapouriser and superheated before reaching the furnace. It will, therefore, be realised that if this highly heated air is allowed to become saturated it will carry forward to the fire a great deal more steam than is necessary or desirable to secure the best quality of gas, and the general efficiency of the producers would thereby be greatly reduced. The significance of this point will be further emphasised when we deal with the design of vapourisers.

7. One of the great advances in gas plant design which has marked the introduction of the suction producer is the utilisation of the heat in the gases leaving the producer for the purpose of raising the necessary steam required for the enrichment of the gas, etc., in accordance with the principles just laid down. On the general reasoning we have adopted it might at first sight appear that if the steam be raised in an independent boiler, a correspondingly larger amount of heat will be available in the producer to allow the endothermic reaction $C + H_2O = CO + H_2$ to take place in a greater degree.

The fact of the matter is, however, that as the general temperature in the furnace of the producer must in any case be from 600° C. to 1000° C. to allow of the decomposition of the fuel at all, the gases leaving the furnace must be at some temperature in the region of those just named whether the steam be supplied from an external source or not. If, therefore, this heat in the gas leaving the producer is not employed for steam raising and for pre-heating the air passing to the furnace, it has to be extracted in the scrubber where it is lost to the cooling water passing through the latter. The investigations of this chapter shew that with a vapouriser and air regenerator of reasonable efficiency more steam can be made from the heat of the gases as they leave the furnace than is actually required or is desirable.

CHAPTER 3.

GENERATOR DESIGN.

It has already been shewn in the general description of a typical suction gas plant given in Chapter 1 that the following are the essential features of all suction generators:

(*a*) The furnace with its firegrate.
(*b*) Vapouriser.
(*c*) Connecting passages between vapouriser and ashpit.
(*d*) Gas outlet from furnace to scrubber.

In the present chapter the various considerations which influence the design of these respective parts will be examined and it will be found possible to proportion all these parts on rational lines, having regard to the general theory of such producers laid down in the previous chapter.

Furnace Design.

Sectional area of furnace and depth of fuel. The French chemist Boudouard has shewn in his experiments on the effect of varying velocities of air currents through heated carbon that generally speaking the higher the velocity the greater the proportion of carbon dioxide produced. Experience has proved this to be true in the case of suction producers, and the whole trend of practice during the last three years has been to make the furnace area much larger for a given duty than was formerly thought to be necessary.

In this latter connection it must be remembered that in all producers which are required to work without stoppage over a working week, the removal of the clinker and ash is practically impossible without the whole fire be drawn and the producer emptied. It is true that by a careful arrangement of cleaning doors and poking holes a proportion of the clinker and ash can be removed while the producer is at work, but even when handled by a skilled attendant well accustomed to the work the greater proportion of the clinker remains in the furnace after each cleaning operation, as it cannot be got at and removed.

It follows, therefore, that the furnace must be sufficiently large to allow the necessary amount of fuel to be reacted on without forcing even though the sectional area be considerably reduced through the presence of clinker at the end of the week's run.

Another important practical consideration is that there is a growing tendency to use cheaper fuels in suction plants. It has been pointed out in Chapter 1, page 17, that gas coke contains a very much larger proportion of ash than Welsh anthracite, and the Scotch anthracites all yield more ash and clinker than the Welsh variety. As gas coke and Scotch anthracite are otherwise suitable for use in these gas plants and are invariably used in Scotland and in many of the northern counties of England, and as the gas plant maker must be prepared to supply plants for use in any part of the country, it follows that the proportions of the generators must be based on the inferior non-bituminous fuels rather than on the best Welsh anthracite.

The published particulars* given below of various makers' plants who took part in the Suction Gas Plant Trials held by the Royal Agricultural Society of England in June, 1906, give authentic information as to the current practice now obtaining for plants of about 20 B.H.P.

It will be seen from Table 9 that the furnace areas adopted by the leading makers work out at about 8 *sq. ins. per* B.H.P. This proportion may be accepted as being well on the safe side providing the arrangements for steam supply are adequate, and in producers of 100 H.P. upwards a proportion of 7 sq. ins. per B.H.P. will give satisfactory results.

Depth of fuel. It will be understood that if the depth of fuel in the furnace be insufficient, a large proportion of carbon dioxide will be produced as complete combustion will take place instead of that partial combustion only which is necessary for yielding carbon monoxide, and consequently the producer will be ineffective from the point of view of producing a burnable gas. In the evolution of the modern suction producer it has

* *Report on the Trials of Suction Gas Plants carried out by the Royal Society of England at the Derby Meeting of* 1906, *by Capt. H. Riall Sankey, R.E., M.Inst.C.E.*

TABLE 9.

Name of plant	Grate area	Cross-section of furnace	Volume of furnace	Volume swept by piston during suction stroke at Catalogue revs.	Ratio of volume swept by piston to		B.H.P. obtained per sq. ft. of grate area		B.H.P. obtained per cub. ft. of furnace volume	
					Grate area	Cross-section of furnace	Full load anthracite	Full load coke	Full load anthracite	Full load coke
	sq. ft.	sq. ft.	cub. ft.	cub. ft. per min.						
National	0·72	1·31	2·37	77·8	109	60	28·6	28·5	8·7	8·6
Railway and General	0·66	1·22	2·44	100·0	151	82	30·3	30·6	8·2	8·3
Davey-Paxman	1·00	1·07	2·28	68·0	68	63	16·4	13·4	7·2	5·9
Dowson...	0·66	1·22	2·44	77·8	118	64	30·9	30·9	8·4	8·4
Campbell	0·79	1·39	3·60	78·0	99	56	24·4	21·5	5·4	4·7
Campbell (Throttle)	0·79	1·39	3·60	87·2	111	63	27·1	21·6	5·9	4·7
Dudbridge	0·79	0·79	1·45	74·1	94	94	19·6	19·7	10·7	10·8
Mersey	0·85	1·07	2·40	66·3	78	62	23·0	19·9	8·1	7·0
Hindley				94·0						
Kynoch	1·00	0·54	0·96	79·6	80	147	18·4	18·0	19·2	18·8
Newton..........................	3·19	0·92	1·84	66·3	21	72	5·7		9·9	
Fielding	Revolving grate	0·66	1·32	81·8		124			14·0	14·2
Crossley	Special grate	Did	not run							
Crossley	Special grate	0·79	1·58	62·6		79			9·7	8·8

been necessary to carefully experiment with generators having different depths of fuel in order to arrive at the correct depth necessary for successful working. From these experiments it has been found that the depth "D" in Fig. 7 should never be less than 18″ in the smallest producers (5 to 10 B.H.P.) and in large producers of 150 B.H.P. and upwards this depth should be not less than 36″. Intermediate sizes may vary pro rata.

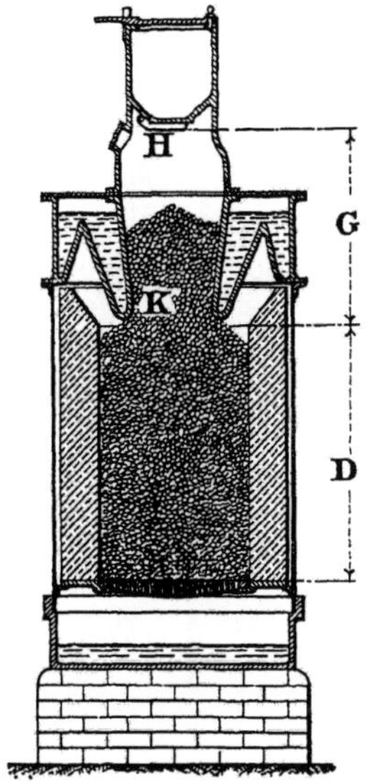

FIG. 7.

Limiting conditions. Experience has shewn that the minimum size of suction producer to admit of successful working must have a lining of not less than 9″ diameter with 18″ effective depth of fuel, and such a producer will give from 5 to 10 B.H.P. In producers smaller than this the working is erratic, due to excessive radiation losses and also to the difficulty in obtaining a sufficiently uniform porosity of fire. In the latter connection the slightest difference in the quality or size of fuel causes great variation in the quality of the gas produced. There does not appear to be any limit in the other direction for suction producers, which have been made successfully up to 500 B.H.P. capacity.

Capacity of coal container. Seeing that the depth "D" in Fig. 7 is the working depth of fuel which is counted upon, it is customary not to allow the fuel to fall below the bottom of the coal container while the plant is at work, hence the capacity of the coal container "G" must be fixed so as to render too frequent firing unnecessary. As a usual proportion, therefore, the container should hold sufficient fuel for a two hours' run at full load, for instance in a 100 H.P. producer the volume of the container between the underside of the valve "H" and its lower extremity "K" should be sufficient to hold at least 200 lbs. of coal, the rate of the consumption being 1 lb. per H.P. per hour. This will mean that if the attendant fills up the container every one and a half hours while the engine is running at full load there will be no fear of the fuel falling below the bottom of the container.

Generally speaking it is an advantage to make the container as large as can be conveniently arranged, and in no case must it be smaller than the capacity just referred to. It must be further understood that where inferior fuels such as French and Belgian anthracites are to be used in suction producers all the proportions of the generators in relation to the power to be developed require to be increased as compared with the practice usual and sufficient where the fuel employed is English anthracite or good gas coke. These considerations are, however, more fully discussed in Chapter 7.

Grate area and firegrates. The area through the firegrate need not be more than one-half the cross-sectional area of the furnace, and where the usual type of firegrate is employed it is found convenient to leave an annular ledge varying from 3″ to 6″ in width according to the size of the plant between the edge of the firegrate and the inner diameter of the firebrick lining of the furnace. Clinker which cannot be got at sufficiently well to be withdrawn from the furnace while the producer is at work may be pushed on to this ledge and the firegrate thus kept free. Some degree of success has been obtained by working suction producers without firegrates at all, and the fuel in the generator simply rests direct on the ashpit bottom as shewn in Fig. 8. Whilst there is much to be said in favour of eliminating firebars and firegrates, it will be found that a grate *does* help to distribute the air supply evenly under all conditions of working, and this fact accounts for the retention of firegrates by most of the leading English makers. In the absence of a grate it is still necessary from time to time to remove the clinker and ash from the fire, and as this clinker forms round the open mouth of the inverted cone "C" which helps to support the fuel above, the withdrawal of the ash and clinker very seriously disturbs the fire and consequently lowers the quality of the gas evolved from the producer. Another objection as to the open grate is that as the

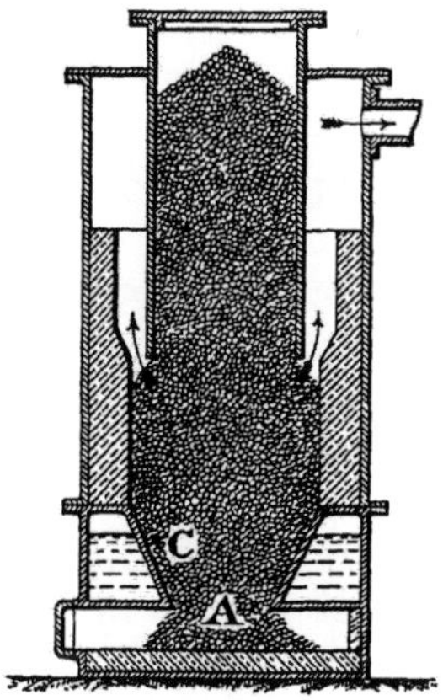

FIG. 8.

inside of the fuel cone "A" becomes very quickly filled with ash and dust if an inferior fuel is being used, the air supply to the furnace is badly distributed, and this has a prejudicial influence on the production of good gas unless the generators are of very liberal dimensions for the required duty.

In plants of 80 H.P. and upwards where it is required to work night and day a week on end, it is necessary to employ some form of moving grate to enable the ashes to be shaken out of the fire from time to time so that the firegrate can be kept free from obstruction for the air and steam to pass through. Details of these moving grates are given subsequently in Chapter 6. When properly designed they are undoubtedly a great acquisition to the working of a large suction plant. They are not, however, required in smaller plants up to 60 H.P. as the cleaning operations in the latter on account of their small size can be dealt with very expeditiously through the fire doors while the plant is still being kept in operation.

It is essential to the preservation of the firebars that an adequate supply of steam be introduced to the fire along with the air supply, otherwise the temperature of the fire will rise to such an extent that the firebars will be burnt away. It follows, therefore, that burnt firebars are almost invariably due to some stoppage of the water feed to the vapouriser, which consequently must be well arranged and properly adjusted.

French makers almost entirely favour the elimination of the firebars and the adoption of open bottoms to the furnaces of their producers as shewn in Fig. 8. As has been already pointed out, however, their conditions are different to ours, due to the inferior quality of the French anthracite, which contains a large percentage of ash and other impurities. Under these altered circumstances good results are no doubt obtained with the open bottom furnaces providing the latter are amply large for their duty.

The vapouriser. It is in connection with the method of utilising the waste heat evolved from the decomposition of the fuel for the purpose of raising the steam which has been shewn to be necessary for the production of good gas, that current practice amongst the various makers differs most widely. The various forms of vapourisers whereby this sensible heat is made to produce the steam can, however, be broadly divided into two classes, namely,

(*a*) *flash boilers* or vapourisers proper, in which no reserve of water is carried, the vapourisation being effected by trickling the water feed over hot surfaces which extract heat from the outgoing gases from the producer;

(*b*) *water carrying boilers* which hold a reserve of water which is in constant contact with the hot surfaces.

Figures 9 and 10 shew sections through the Dowson and Crossley vapourisers which respectively represent the two types referred to above, and, to further emphasise the distinctive difference between them, we give in Figs. 11 and 12 sectional elevations of the latest National and Tangye vapourisers.

The question of the relative merits of these two types of vapourisers naturally arises, and apart from practical considerations of construction and workmanship, it is desirable

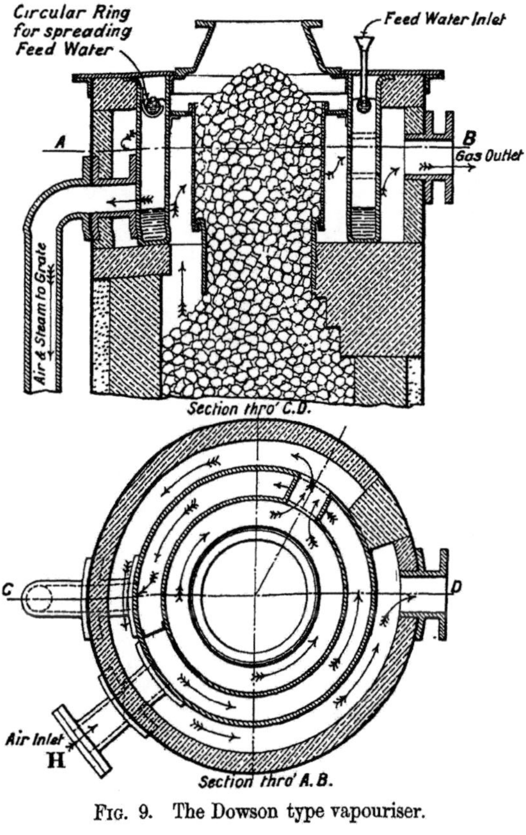

FIG. 9. The Dowson type vapouriser.

in the first place to enquire if the working results obtained with the two types differ sufficiently to warrant the choice of one in preference to the other. It is to be admitted that there are a great many of both types of plants successfully at work, but the author considers that the flash boiler type of vapouriser is best calculated to serve the end in view, and so far as any impartial comparative tests have been carried out, plants with this type of vapouriser have always excelled those with the alternative type. There are many reasons why this should be so, as the following considerations will shew.

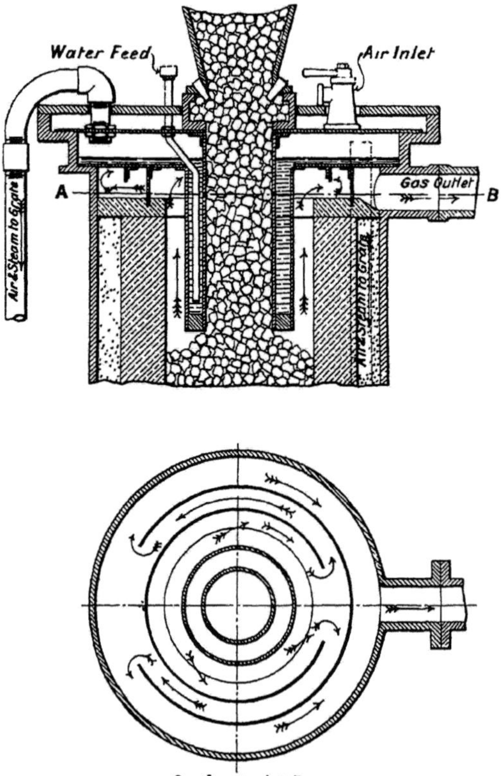

FIG. 10. The Crossley type vapouriser.

In all plants the temperature of the air is raised considerably before it passes to the fire, and consequently if this heated air is allowed to be in contact with an unlimited supply of water vapour it will carry forward to the fire an amount of steam in accordance with known rules as to the weight of water in each pound of air at saturation point for the temperature of the ingoing air in question.

In Fig. 13 a curve is set out shewing the weight of water per pound of air at saturation for different temperatures, and the shape of the curve shews how rapidly the weight of water which can be carried increases as the temperature rises.

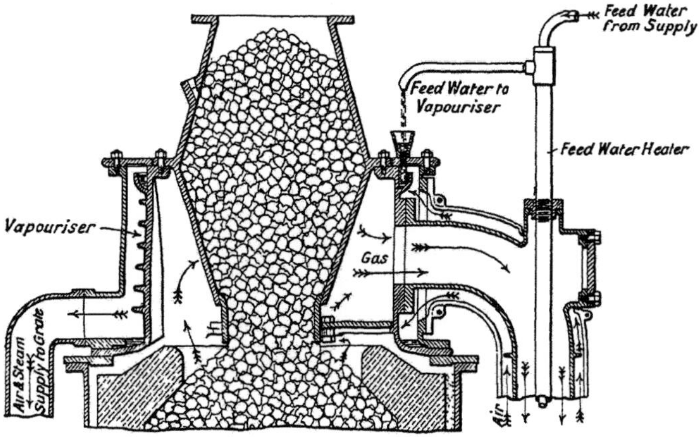

FIG. 11. The National type vapouriser.

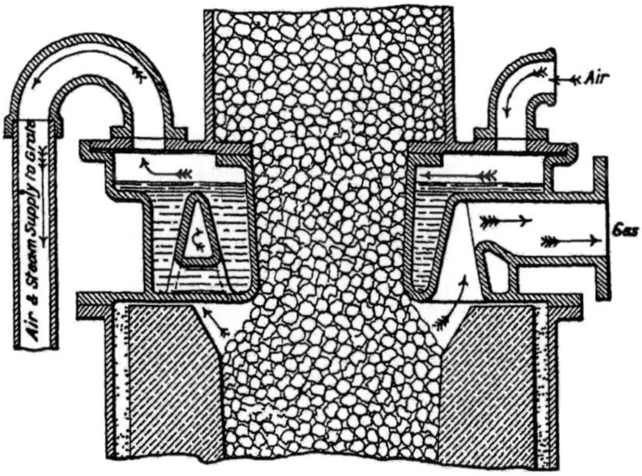

FIG. 12. The Tangye type vapouriser.

Looked at from another point of view this means that the heat units contained in one pound of saturated air, together with the moisture it contains, also rise very rapidly as the temperature increases, and Capt. Sankey in his report on the Derby trials prepared a curve which is given in Fig. 14 to draw attention to this fact.

It has been argued therefore that, by raising the temperature of the air passing to the furnace to as near 212° F. as possible, and allowing this air to be fully saturated, a much greater recovery of the waste heat would be possible, and also a

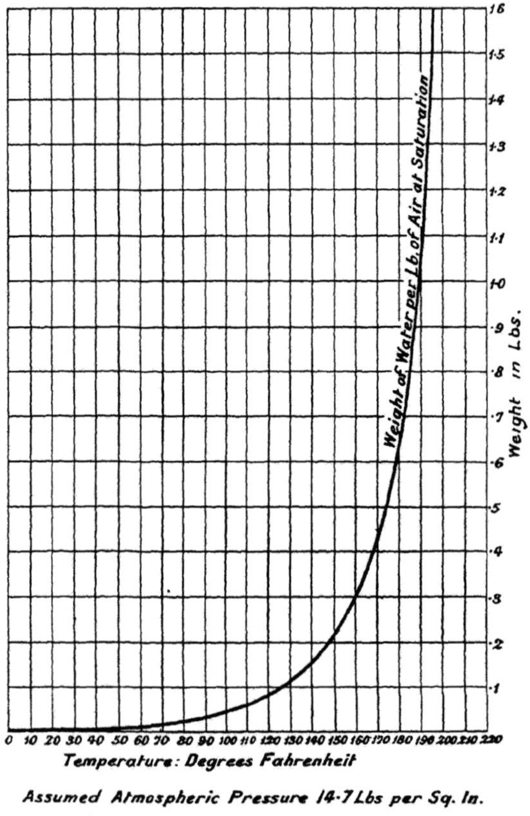

FIG. 13. Curve shewing weight of water carried per pound of air at different saturation temperatures.

consequently higher general efficiency would result. There is however a serious fallacy in this argument which cannot be sufficiently emphasised, for it assumes that the weight of steam which can be carried by saturated air at any temperature up to

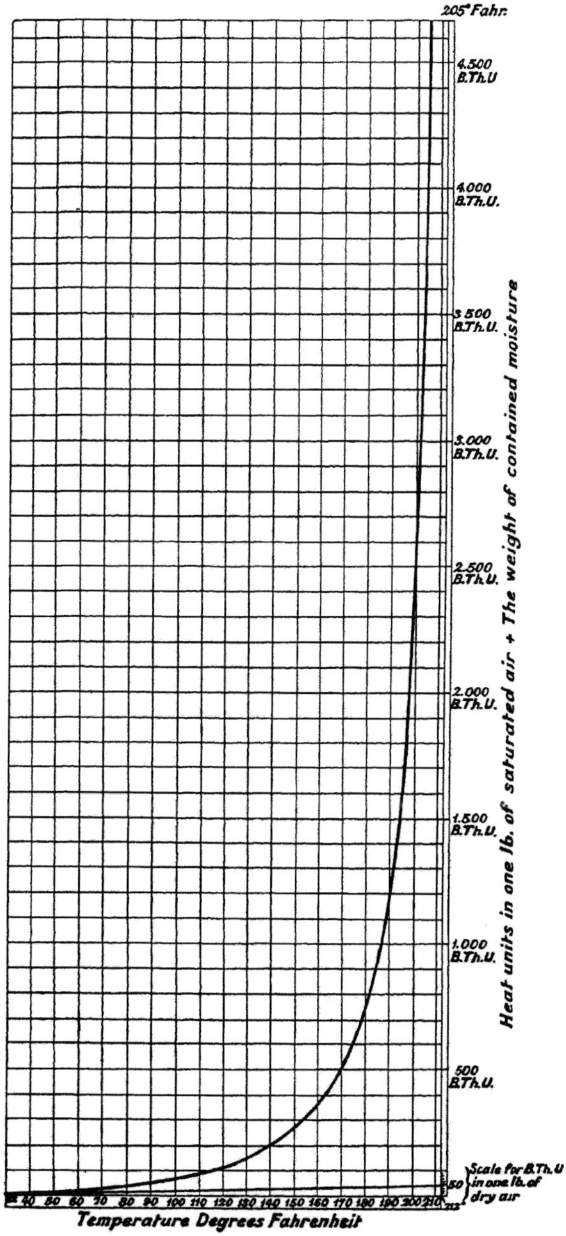

FIG. 14. Curve of Heat Units in saturated air at different temperatures.

212° F. will be decomposed in the fire under conditions to give good results. This is not the case for the reasons given in Chapter 2, page 30, for it is there clearly shewn that the correct amount of steam to give the best results corresponds to that which can be carried by fully saturated air at a temperature of about 85° F. only. Saturation at higher temperatures therefore means that too much steam is being carried forward, and as a result there is great loss of efficiency and reduction in the heating value of the gas produced. It must be borne in mind that it is quite correct to heat the ingoing air to as great a degree as possible so as to enable it to carry forward the maximum amount of steam to be decomposed without unduly *lowering the temperature of the fire*, but the air at any temperature above 85° F. should not be fully saturated. The important difference between the action of the flash boiler as against the water container type of vapouriser is that in the former it is possible to regulate the degree of saturation of the ingoing air with accuracy to give the best results, whilst in the latter, as the air has to pass over a large surface of steaming water, the same control of the steam cannot be approached.

TABLE 10.

Rate of Water used in Vapourisers.—Full Load Coke Trial.

Name of producer	Observation. 500 cub. centims. in:		Rate in gallons	
			Per hour for plant	Per B.H.P. per hour
	min.	sec.		
National	3	16	2·0	0·1
Kynoch	1	11	5·6	0·32
Fielding	4	32	1·5	0·08
Crossley	0	58	6·8	0·5

In practical support of these statements the author has frequently caused an engine working from a National suction producer on absolute full load to lose speed merely by increasing the water feed to the vapouriser and thus increasing the steam supply to the fire. This slowing down is, of course, caused by a falling off of the quality of the gas due to excessive steam, but if the air supply to the producer had been fully saturated in the first instance no further increase in the steam supply to the fire would have been possible. The correct amount of steam to give the best results could not be regulated in the same way with a boiler containing a large volume of water delivering steam to the incoming air *ad lib.*, and this point may be further demonstrated by a comparison of the published figures given in the official report on the Suction Gas Plant Trials held by the Royal Agricultural Society in June, 1906, which are given in Table 10.

It will be seen from this table that the Crossley plant required a rate of feed to the vapouriser five times as great as in the National plant. Though the Crossley feed is undoubtedly in excess of actual requirements, the design of their vapouriser with its extensive heating surfaces and the large body of water contained would appear to have the tendency to produce a larger volume of steam than is necessary or desirable, and as a consequence their arrangement of primary and secondary air supply described on page 56 is introduced. It is of course better to be without these additional complications, and with the alternative form of vapouriser referred to they are not necessary.

The further figures given below from Table XII of the report referred to shew conclusively moreover that the gas produced with this excess of steam could not have been of such a high calorific value as that evolved in the Dowson or National type of producer.

TABLE 11.

	Maximum initial pressure in lbs. per sq. in.	Compression pressure in lbs. per sq. in.	Percentage "hits"	B.H.P. per cub. ft. swept by piston
National......	240 to 390	160	85	0·131
Crossley	435 to 510	210	93	0·123

These differences are further emphasised by a comparison of the respective indicator diagrams which are reproduced in Fig. 15 on the next page.

It will be seen that though the Crossley engine was relatively larger for its work, it had a smaller margin of power available, whilst although the higher compression on this engine should have ensured a reduced consumption of fuel, the actual consumptions at full load were 1·04 pounds per B.H.P. per hour with the National engine as against 1·15 pounds per hour with the Crossley engine.

It will be seen, therefore, that the flash boiler type of vapouriser gives a better control of the steam supply, and instead of absorbing too much heat and producing needless steam it ensures a ready means of vapourising only the correct amount of water, leaving any surplus heat for the purpose of

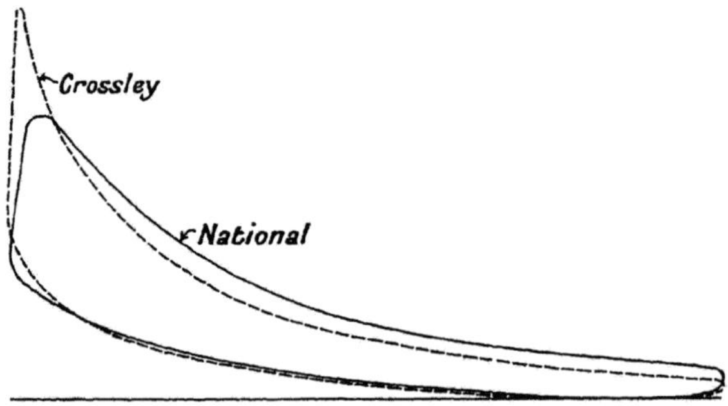

FIG. 15. Indicator Diagram from National Engine to Full Lines.
" " " Crossley " " Dotted "

superheating the steam or preheating the air. The preheating of the air naturally enables the maximum amount of steam to be used in the fire with due advantage to the quality of the gas made, and hence, in our view, it enables the highest efficiencies to be obtained by thoroughly practical means.

Position of vapouriser. In modern practice the vapouriser is almost invariably placed at the top of the generator. Producers have been made, and are being made now, by a few firms with the vapouriser in the form of a boiler encircling the generator close to the hot zone of the furnace as shewn approximately in Fig. 16, but this practice has little to commend it.

If the gas yield is to be automatically regulated to suit the load requirements the temperature of the fire should be entirely dependent on the amount of ingoing air and steam. With a boiler or vapouriser in the position shewn in Fig. 16, however, the effect must clearly be to extract too much heat from the fire so that at full load too much steam is made, which must be wasted unless an overdose of steam is to be given to the fire.

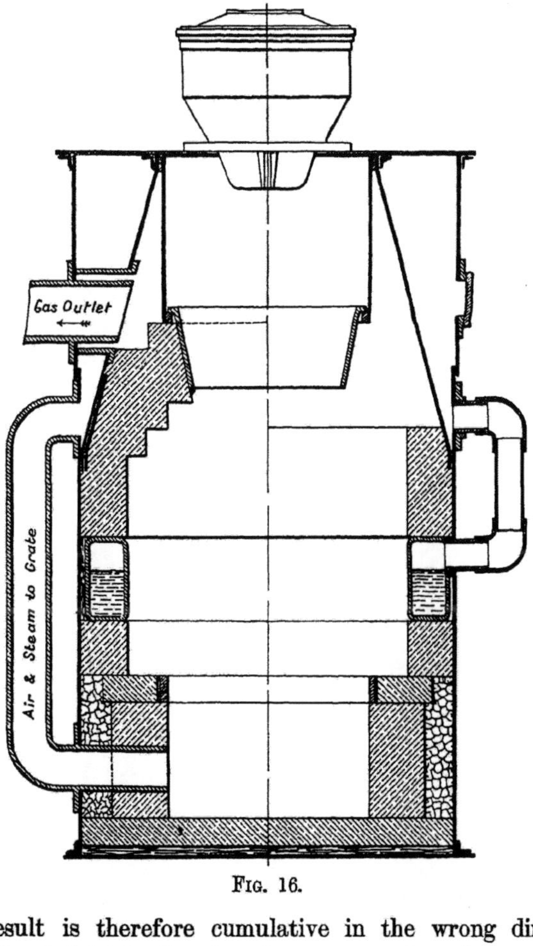

FIG. 16.

The result is therefore cumulative in the wrong direction. Similarly at light loads, when in any case there is a tendency for the fire to die down, the effect of a boiler in this position is to chill out the fire altogether. It will, therefore, be seen that the proper place for the vapouriser is at the top of the producer, where it extracts heat from the outgoing gases without interfering with the ordinary temperature of the fire, which is thereby regulated entirely by the amount of air and steam drawn in while the producer is at work.

In the Kynoch producer shewn in Fig. 17 the vapouriser is separated from the generator body altogether, and while this enables a simple plan of construction to be followed, it is difficult, in all cases, to arrange for the same amount of heating surface to the vapouriser which can be conveniently obtained when the latter is embodied in the top of the generator. Still, if properly worked out, the design can be made successful.

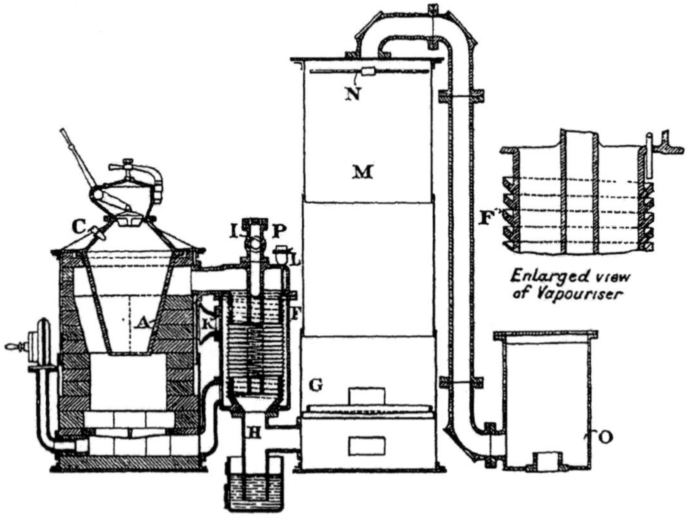

Fig. 17. Kynoch Producer.

Vapouriser heating surface. So far as the proportions of heating surface to the vapouriser are concerned, seeing that the outer diameter of the generator body is fixed by the sectional area required for the furnace and that the necessary capacity of coal container controls the total height of the plant above the furnace, it is usual to work in the vapouriser, which is on the outside of the coal container, to suit the diameter of the generator and the height of the container. The proportions thus obtained are found to answer well in practice.

Automatic control of the water feed to vapouriser or steam supply to the fire. It will have been already noted that it is in the design and arrangement of the vapouriser that the practice of various makers differs most, and there have also been great differences of opinion from time to time as to the need of providing means so as to automatically regulate the amount of water or steam feed so that a definite quantity of the latter is arranged to be injected or supplied as the case may be for each suction stroke of the engine. The necessity for such means has been thought to arise from the difficulties which are sometimes experienced in working on a widely fluctuating load. For instance, supposing a plant with flash boiler vapouriser has been working for some time on full load so that the temperature of the fire is raised to its maximum extent and the load then suddenly falls off, two conditions appear to follow. If the water feed to the vapouriser continues at the former rate suitable for full load an excess of steam in proportion to air is made until the vapouriser is cooled down, and this steam passing through the hot fire (which has been left at the maximum temperature whilst the engine was running at full load) causes water gas to be produced for a short period afterwards, and when this comes through to the engine, the gas cocks of which are adjusted for the ordinary conditions, excessive initial pressures are produced in the working cylinder. The excess of steam has the further effect of unduly cooling the fire, so that if a full load demand is made on the producer shortly afterwards it is unable to meet it, with the result that the engine tends to stop. The author has found, however, that the production of water gas under these particular conditions is due far more to the reduced velocity of the saturated air passing through the fire when the load is reduced than to the water feed to the vapouriser being allowed to continue at a constant rate, for it will be found that even when the water is shut off altogether simultaneously with the removal of the load, the water gas is produced practically to the same extent as before.

When the plant is called upon to meet a large increase of load the automatic regulation of the water feed does not help matters at all. The actual behaviour of suction plants under changing loads is dealt with more fully in Chapter 10, but it will be appreciated here that *prima facie* if the water or steam supply is made automatic with the suction strokes of the engine as soon as the load comes on and before the fire has been brought into a good condition, the maximum water supply is being drawn in and in this respect the regulation is no better than that obtained with a constant feed regulated in such a way as to suit the average load. These considerations, together with the practical results obtained, have confirmed us in the opinion that automatic regulation of water feed is an unnecessary complication which confers no benefit and which consequently may be safely avoided.

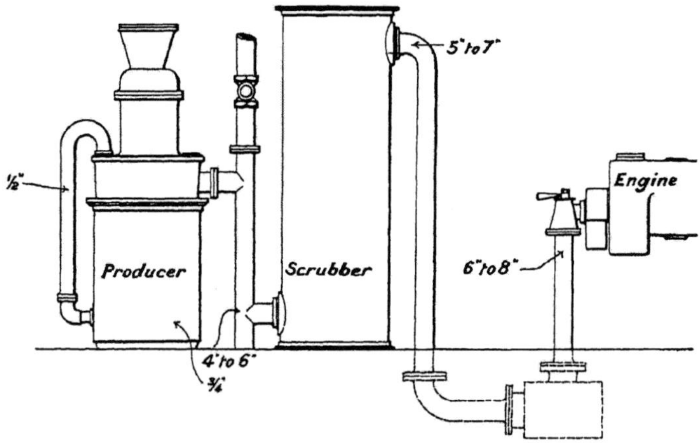

Fig. 18. Diagram of negative air pressures at different points of suction plant during suction stroke of engine.

From another point of view it is very anomalous that so much attention should have been paid to secure an automatic regulation of the water supply in accordance with the power strokes of the engine which, for the reasons given above, hardly matters at all, and that so little effort should have been directed by the majority of makers to devise means to ensure that correct proportion of steam and air in the supply of the latter to the furnace, which is vital to efficiency.

Area of inlet and outlet passages for air and vapour and gas. The actual sizes of these passages require to be determined from actual experience with each type of plant, but they must in every case be sufficiently ample to prevent excessive back pressure against the suction action of the engine. Reasonable suction pressures as determined by a water gauge in inches of water at various points of the plant are given in Fig. 18. Any impediment or choking effect at any part of the plant can always be located by a logical comparison of these suction readings at various points.

CHAPTER 4.

CONSTRUCTION OF GENERATORS.

Generator Body. The practice of various makers in the selection of the material for constructing the generator bodies varies considerably. The choice lies in the use of cast iron as against wrought iron parts. When plants were first introduced there was considerable doubt in the minds of the makers as to whether cast iron could be safely used at all on account of the very considerable temperatures to which the generators are subjected. On the other hand, the desirability of securing an absence of corrosion, together with the importance of having machined joints so far as possible, make cast iron parts preferable wherever they can be introduced for this class of work. As a result of actual experience with suction plants since their introduction the author is of opinion that cast iron may be safely used in constructing generators up to 100 B.H.P., and this practice is now being followed by most leading makers, including Messrs Crossley Bros., Tangyes, and the National Co. For plants of 150 B.H.P. and above the cylindrical outside shell of the generator can be more conveniently constructed in wrought iron plates riveted together, but in every case the vapouriser and ashpit should be made in cast iron, for these parts are subjected to the continued action of saturated air which tends to produce excessive corrosion. We have actually known plants constructed of wrought iron throughout in which the plates forming the vapouriser and ashpit have corroded into holes within two years.

Importance of good joints. The importance of having good joints throughout the plant cannot be sufficiently emphasised, and there is not the slightest doubt that failure to realise this point has caused many makers a great deal of trouble. This is particularly the case with joints in the upper portion of the plant. It will of course be borne in mind that seeing the plant works on the suction principle the tendency is for air to leak *in* and not for gas to leak out. Any air leaking into the top of the plant causes the outgoing gas to ignite or to partially

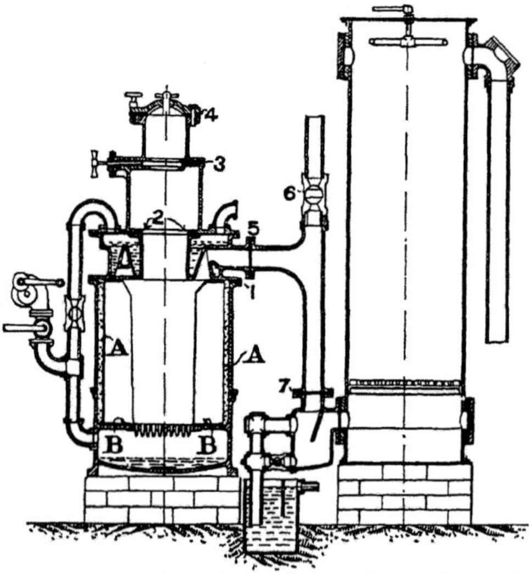

FIG. 19. Section through suction plant: the figured joints shew where air leakages are specially detrimental.

ignite, and this burning of the gas as it comes off naturally makes it of little use for the engine afterwards, and in addition the heat evolved in the upper part of the plant from this cause is quite sufficient to burn away the plates if of wrought iron, and to crack the vapouriser and outlet pipes if of cast iron. It is also important that no serious leakage of air takes place through the joints in the firebrick lining, and in short the successful action of the plant depends primarily on the air being admitted at the proper place only, namely, through the firegrate. In Figs. 19 and 20 we have marked the particular joints which must be carefully made, and have also indicated where no air leakage can be allowed, and makers and users of gas plants will do well to carefully follow our instructions in this respect.

Firebrick Lining. The firebrick lining of the generator should be a thoroughly substantial job, and its thickness should be not less than 3 inches for the smaller size of generator (5 to 10 B.H.P.) to 8 inches in the larger generators (100 B.H.P. and upwards).

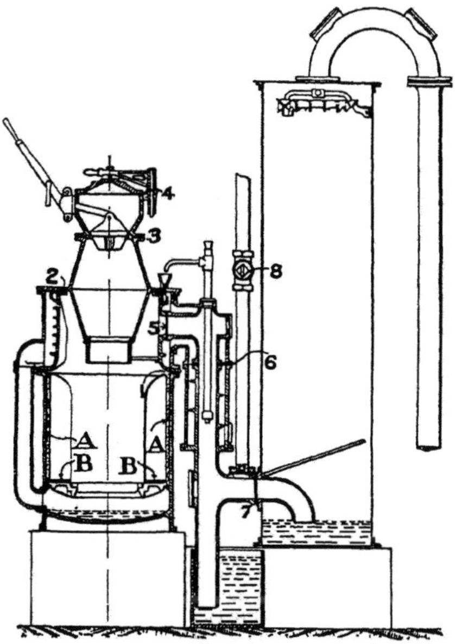

FIG. 20. Section through suction plant: the figured joints shew where air leakages are specially detrimental.

There are several points in connection with these firebrick linings which must be carefully appreciated. In the first place it will be realised that it is a practical impossibility to make the outer diameter of the lining a perfect fit to the inner diameter of the generator body, and it has become the standard practice to allow a definite space or annulus A.A. (Figs. 19 and 20) all round between the brick work and the latter. This allows the brick work to expand or contract irrespective of the generator body as it heats up or cools down. Under the working conditions, however, it will be understood that there is a constant tendency for the joints in the brickwork to open out, and apart from these joints cracks frequently appear in the bricks themselves at the lower part of the furnace where the heat is most intense. Further to this it is impossible to make a satisfactory joint between the lower surface of the brickwork B. B. which rests on the supporting ledge of the generator body, capable of permanently preventing leakage of air from the ashpit through this joint into the annular space A. A., and from thence, in the absence of preventive means, through the seams of the brickwork into the furnace. The danger of such a leakage is to set up a zone of combustion in the upper part of the furnace where its presence may completely vitiate the successful working of the producer, as the inflammable constituents of the outgoing gas, which would otherwise pass off for use in the engine, are burnt to carbon dioxide by this leaking air before leaving the producer. The heat set up at the top of the generator when this occurs, as has already been pointed out, is sufficient to ruin the vapouriser and outlet pipes if allowed to continue, whilst, of course, the gas produced, having already been consumed or partially consumed, is useless for its purpose.

It therefore follows that too much care cannot be taken in setting up the firebrick lining in the first instance, and as each joint in the brickwork is a possible source of leakage, the number of such joints should be reduced to a minimum. Specially shaped bricks must be made for each size of generator and the radial joints A. A., Fig. 21, should be a thoroughly good fit, so that when laid to form a complete circle they butt evenly from the inner to the outer diameter with only a slight space between for a very thin layer of fireclay jointing. A suitable arrangement of lining for a 80 H.P. plant is shewn in Fig. 21.

To prevent leakage of air through or up the annular space A. A., Fig. 19, between brickwork and generator body it is usual to pack it with sand or fireclay. Of these two materials the former is to be preferred as it naturally shakes down and becomes automatically adjusted to any change of shape in the brick or iron work. The only objection to it is that if put in place before the plant is despatched from the makers' works it is apt to get shaken out altogether in the vicissitudes of transit. The author has used fine silicate of cotton with most satisfactory results, for this material when packed tight has not the disadvantage just referred to and is besides a good non-conductor of heat.

It is most important that the bricks should be of the very best quality, for the use of soft or unreliable material will only cause a great deal of annoyance and loss to both the gas plant maker and the user. Excepting as an emergency repair when nothing

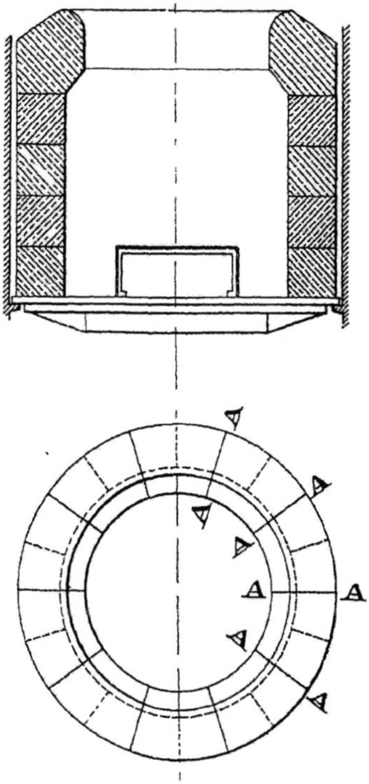

Fig. 21. Diagram of Furnace Lining.

better can be obtained, no attempt should be made to build up the furnace lining of these plants with bricks other than those specially shaped for the particular size of plant, for if the bricks have to be chipped to shape the cut surface is both rough and soft and thus encourages the formation of clinker when the plant is at work, and is less able to resist, without damage to the lining, the detachment of this clinker afterwards.

Current Practice. The Figs. 22, 23, 24, 25 and 26 give sections of generators in accordance with the practice at present adopted by Messrs Crossley, Dowson, Tangye, Campbell, and the National Company respectively in their suction plants, and though there are numerous other makers each of whom embody some special feature of more or less importance in their productions, it is thought that those of chief repute who have been selected will sufficiently represent the best modern practice.

Messrs Crossley Bros., Fig. 22. This illustration is in accordance with the published information respecting the plant exhibited at the Derby Trials by this firm as their standard plant of 15 B.H.P., and we believe the general type is followed closely for all the sizes they make. The generator body is a single shell of cast iron in one piece from the base to the top joint: in the larger sizes, however, we believe this shell is built of steel plates. The heating surface of the vapouriser is formed of steel plates bent into a concentric ring of U section to form the lower neck of the coal container. The outgoing gas passes from the top of the fuel up the sides of this ring and is then caused to move over the projecting flat surface of the vapouriser by having to pass through passages A, B, formed by the guiding plates attached to the vapouriser and projecting into the upper surface of the brickwork. The steel part of the vapouriser is secured to the inner facing of the bell-mouth top of the generator shell, which is recessed on its top edge to carry a cover plate over the vapouriser.

An ordinary cock is fitted to the cover plate as an air inlet, which is known as the secondary air supply, and there is both an external pipe C and an internal tube D for conveying the saturated air from the vapouriser to the fire. This design of vapouriser would appear to have the tendency to produce a liberal supply of steam, and as under working conditions this is frequently found to be excessive, a damper is fitted to the air inlet E from the fan, which is used as a primary air supply. This latter is opened when the steam supply is found to be excessive, in which case the secondary air inlet acts as a relief opening from which surplus steam escapes. According as the primary air damper is regulated, more or less air will be supplied to the

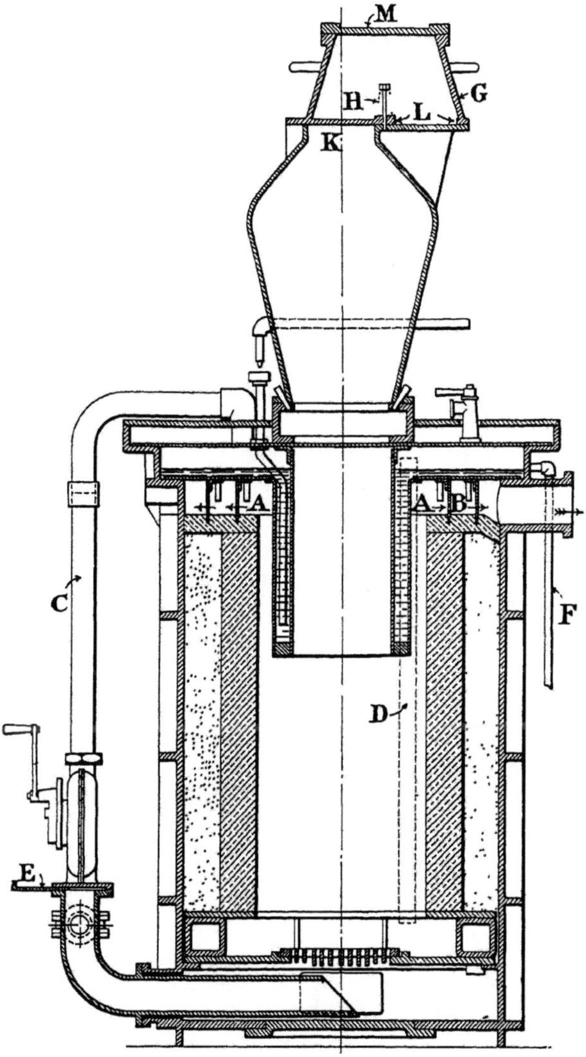

FIG. 22. Sectional elevation of Crossley suction generator.

fire from this source as compared with the secondary air inlet, and consequently the steam supply to the fire is within certain limits under control. A further method of regulation is provided by the external pipe C, through which steam may be introduced to the primary air direct.

This arrangement of dual air supply would appear to us to necessitate considerable judgment and experience on the part of the attendant if it is to be of real service, and other makers successfully avoid this difficulty to a very large extent by having the production of the steam in the first instance under effective control.

The coal container proper is of cast iron secured as shewn in Fig. 22. Poking holes are provided at its base so that a poker may be passed in and forced through the fire if an arch of clinker becomes formed whilst the plant is at work. Opinion is divided amongst experts as to whether these are necessary or not in a well designed plant. It is certain, however, that an anthracite fire should be disturbed as little as possible, and when this fuel is of the usual quality poking will not require to be done while the producer is working, if the generator is sufficiently large for its work and if the steam and air supply is properly adjusted.

. A special feature of the arrangement of the furnace is the introduction of the cast iron square section concentric tube between the grate and the firebrick lining. This tube or superheater, as it is called, is made in two half-circular pieces, and from its position it naturally catches a great deal of heat from the fire, but is prevented from becoming overheated since the air and steam from the vapouriser are made to circulate through before passing beneath the grate. Any clinker which is formed tends to drop to the bottom of the fire and in the ordinary way adheres to the firebrick, from which it gradually grows, but it is claimed for the design under consideration that it prevents both the adhering and the growing referred to. The Judges at the Derby Trials commented favourably on the arrangement, though it is open to doubt whether the freedom from clinker on the occasion of the Trials was not more probably due to the excess of steam used (see Chap. 3, page 44), which, though favourable to the non-production of clinker, has the corresponding prejudicial effect on the quality of the gas produced.

The water supply to the vapouriser is introduced through an open funnel connecting with an internal pipe which carries the cold feed to the lowest part of the vapouriser. The regulating cock on the supply pipe is set so that sufficient water flows into the vapouriser to keep a constant overflow through the run off pipe F, and the water level in the vapouriser is thus kept at a constant height.

The arrangements provided for feeding the fuel are somewhat different from those usually adopted. The coal hopper valve G revolves on the spindle H: the hole K in the coal container is normally shut off from the hole L in the hopper valve. The operation of feeding in a fresh charge of fuel consists in firstly swinging back the top cover plate M and introducing the fuel into the hopper valve, after which the top cover plate is replaced. The hopper valve is then rotated by the side handles until the two holes K and L are opposite to each other, when the fuel drops through into the producer. The hopper valve is then rotated back into the first position and the same operations are repeated until the producer is charged up to the top. It is of course necessary in all suction plants to provide means for introducing the fuel without air being allowed to pass into the top of the producer and a double valve arrangement of some kind must be adopted.

The Dowson Plant, Fig. 23. These plants have always been noted for their high efficiency, and in their general design sound theoretical principles are embodied. As will be seen from Fig. 23 the generator consists of a single outer shell with fire-brick lining and ashpit arranged as shewn. The mild steel vapouriser A acts on the flash boiler principle and is circular in form and of U section. The vapouriser cover plates B are arranged in segments which can be easily removed, and the water feed is supplied through an external open funnel to the circular ring C and from thence through a series of nipples on to drip plates D, which spread it over the heated sides of the vapouriser. The outgoing heated gases, after issuing from the top of the fuel, are firstly caused to circulate round the passage E between the inner sides of the vapouriser and the coal container, and thence pass through to the outer passage F between the

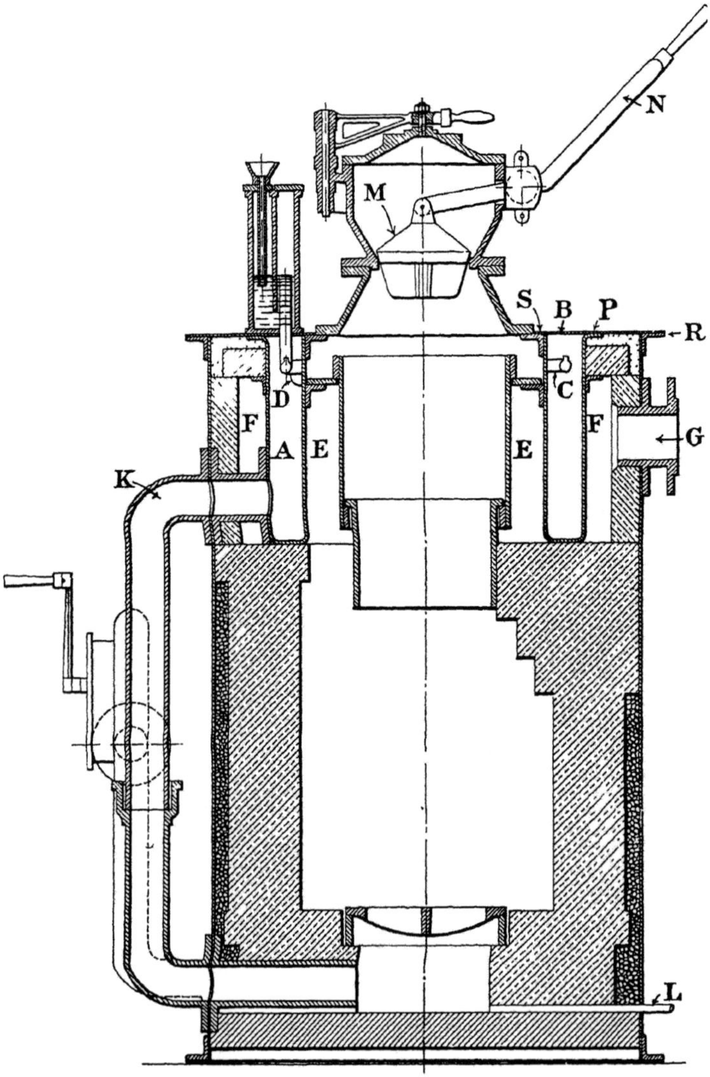

FIG. 23. Sectional elevation of Dowson suction generator.

other side of the vapouriser and the outer shell: the outlet pipe G to the scrubber connects to the passage F.

The incoming air passes to the vapouriser at the inlet branch H (see Fig. 9) and a division plate causes it to circulate round the vapouriser before issuing to the air and steam supply pipe K, by which the saturated air is conveyed from the vapouriser to the grate. The height of the outlet branch from the vapouriser is arranged at only a short distance above the bottom of the latter and consequently if the water feed be in excess of the steam made the surplus water flows down the connecting pipe G to the underside of the grate, from whence it is drained by the run off pipe L. It is not intended that this vapouriser shall carry a large volume of water as in the Crossley, Tangye, and Campbell types of vapouriser.

The hopper valve arrangement by which the fuel is introduced is simple and convenient. The usual form of hinged top cover plate is used, and the inner valve M is a drop valve operated by the external lever N. This drop valve has a conical seating, the angle of which differs from that on the fixed seating, and this makes it easy to bring the valve down tight even when particles of coal are lying thereon. There is nothing to get out of gear, and the fact that the valve and seating can be turned to suit each other in the first instance makes it possible to secure a permanently satisfactory job. Excepting for the tendency to corrosion which exists in all plants made wholly of wrought iron and the necessity for very careful workmanship on the top joints P, R and S, which in wrought iron are difficult to bring straight in the ordinary way, these plants are a good job and their working capabilities are satisfactory.

The Tangye producer, Fig. 24. Messrs Tangye were among the first in this country to put Suction plants on the market, and their producer being simple in action and of good workmanship and material has met with a very large measure of success. In the range of sizes up to 120 H.P. the generator bodies are made in cast iron. The shell proper of the generator is made in two pieces A and B which are jointed together by a spigot joint. This plan facilitates moulding and likewise is considered to give more elasticity to these castings under the influence of heat. The vapouriser C is bolted to the top of the generator shell, and the construction in cast iron enables a satisfactory machined joint to be formed at D. The cover plate E of the vapouriser is combined with the coal container. It will be seen that the vapouriser is of the form which carries a body of water, and when the producer is at work air is drawn in at the inlet H, whence it is drawn across the surface of the heated water where it takes up and becomes charged with steam or vapour. A mixture of air and steam then passes down the pipe K to the ashpit, being afterwards drawn through

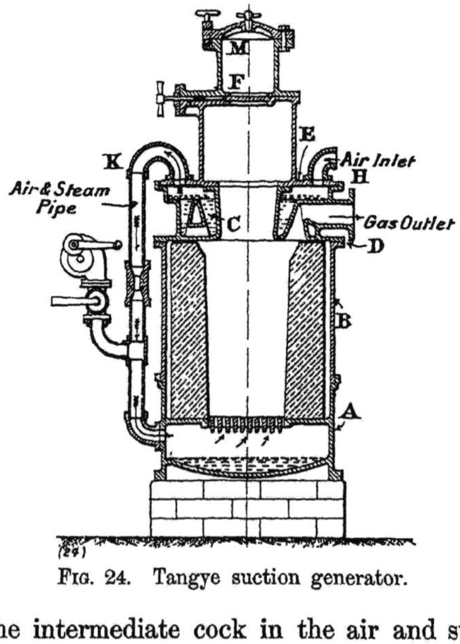

FIG. 24. Tangye suction generator.

the fire. The intermediate cock in the air and steam pipe is provided for the purpose of shutting off the vapouriser when the hand blower is being used to start the plant. The firebrick lining is a thoroughly good job and is made up with specially shaped bricks with the minimum number of joints. The vapouriser is kept nearly full of water by means of supply and overflow pipes: the overflow is carried down and runs into the bottom of the generator, where it gives off additional steam or vapour produced by heat radiated from the hot fuel and also from the hot ashes falling into the water. Any excess of water is carried away by a drain pipe connecting with the water seal pot. The presence of water in the ashpit is very useful when the plant is first being started, for in vapourisers of this type it is some time before the water therein becomes sufficiently hot to give off vapour. Meanwhile, however, the lower part of the plant attains considerable heat, and consequently the water in the ashpit must be relied upon to give the necessary steam supply.

The arrangement of valves in the coal hopper is different from that adopted in the two plants previously described. The inner valve F is a flat plate which is drawn away horizontally, and if poking of the fire is necessary in a plant, this enables the poking bar to be put in at the plug hole M and pushed right down the centre of the fire. As we have pointed out, however, if the plant is of ample size for its work and rightly adjusted in other ways, very little poking is required.

Messrs Tangyes have many hundreds of plants working well. Their design has a neat and workmanlike appearance, and there is no doubt in our mind that the realisation by this firm in the first instance of the importance of good joints, and a first class brick lining, has contributed very largely to the practical success obtained with this type of plant.

The Campbell plant, Fig. 25. In general form the generator of this plant is similar to the Tangye producer just described. The construction, however, is different in so far as the body is built of mild steel plates socket-jointed into the cast iron ashpit. The vapouriser is of cast iron similar in form and action to the previous one, but the principle of regeneration is carried further by causing the outgoing gases to pass over the air and steam supply pipe connecting the vapouriser to the underside of the grate. This is arranged in the so-called superheater, which consists of an external shell A with an internal pipe B through which the air and steam pass from the vapouriser: a division plate C ensures that the outgoing gases from the producer shall pass down the whole depth of the superheater before returning to the outlet to the scrubber, so as to make the heating of the internal pipe fully effective. The general construction will otherwise be easily grasped from the figure.

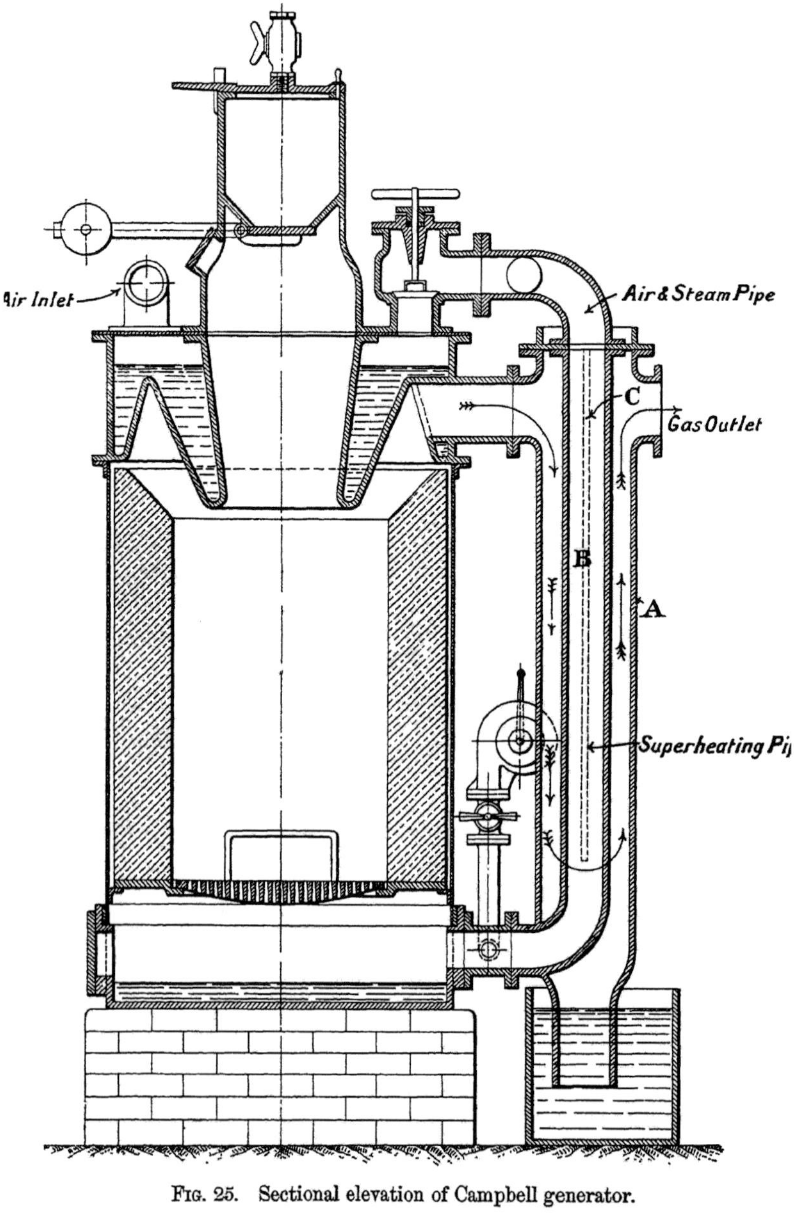

Fig. 25. Sectional elevation of Campbell generator.

It is impossible to say what the gain in efficiency actually is due to this method of superheating, but the principle is correct and the performance of the Campbell plants at the Derby Trials was good.

The hopper valve is again of a different type and in this case is a flap valve held to its seating by an external counterweight on the operating lever. When there is no coal immediately underneath, it can be moved sufficiently clear of its seating to enable a poking bar to be got through, for which purpose a cock is provided in the top cover plate.

Capt. Sankey in his report on the Trials referred to describes this plant generally as being of simple solid construction.

The National plant, Fig. 26. The National plant which was awarded the Gold Medal at the Derby Trials has a generator considerably different in design and construction from that of the other makers. The general action of this type of plant has already been described (see Chap. 1, p. 8) and, from the constructional point we are now considering, it may be explained that the object in view is to secure the maximum efficiency with a simple and reliable design which admits firstly of a sound job, and secondly of modern manufacturing methods of production. Both of these considerations directly affect the user, the former for obvious reasons and the latter because it enables spare or replace parts to be ordered which, being made to proper templets, can be relied upon to fit without trouble when obtained.

The generator body is of cast iron with machined joints throughout for all sizes up to 120 B.H.P. The vapouriser is on the flash boiler principle with steam production under effective control, and in order to equalise the temperature of the various parts of the vapouriser shells, and to secure high efficiency, both the air supply and water feed are preheated before being introduced to the vapouriser.

After being heated the water supply is brought to an open funnel, whence it drops into the trough which is led all round the heated internal shell of the vapouriser. A constant overflow from this trough through V notches is arranged at intervals round the circle, and each overflowing stream trickles down the side of the vapouriser, spreads on the horizontal ribs below, and before reaching the bottom is completely vapourised. If, through inadvertence, the rate of water feed is in excess of the rate of vapourisation, surplus water passes out of the vapouriser down the air and steam pipe to the underside of the grate, whence it is drained off by a pipe provided for the purpose. The heated incoming air from the jacket round the gas outlet pipe catches up this steam and becomes further heated on its way to the air and steam connecting pipe to the grate referred to.

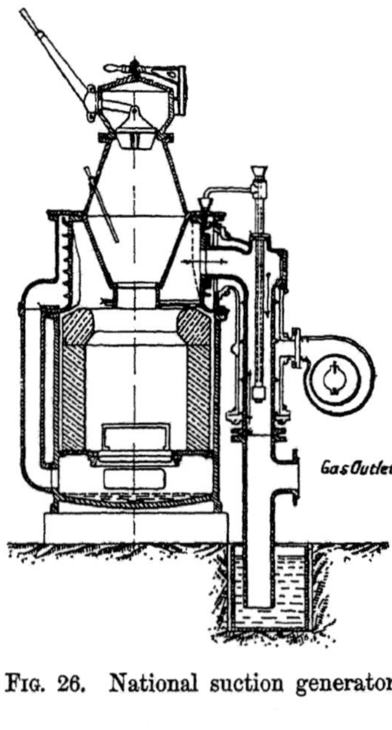

FIG. 26. National suction generator.

It will be observed that the inner shell of the vapouriser, which is the only cast iron part of the generator, excepting outlet pipe subjected to the direct action of the heated gases, is a simple casting of single thickness which therefore precludes any initial casting strains and is moreover free to expand and contract under working conditions without restraint. The design affords another great advantage: if hard or salty water which favours the formation of scale is alone available for use in the plant, the outer shell of the vapouriser can conveniently be removed at any time and thus any deposit can be removed without trouble.

The starting fan is connected direct to the air jacket and thus when starting up air is blown through the vapouriser to the grate carrying forward any steam which may be formed. This is of great advantage where plants have to be stopped and started several times a day and as there is a growing tendency to shut down plants at meal hours in factories, and as in many cases it is absolutely necessary to do so, this method of connecting the starting fan is greatly to be preferred. It enables gas to be got quickly and prevents passages being blown through the fire caused by the forcing action of dry air at pressure.

The firebrick lining is carefully arranged to give the minimum number of joints and the top layer is formed of specially shaped bricks so as to protect the bottom of the vapouriser should the fire arch up at any time through a bad quality of fuel or from other exceptional causes.

It will be observed that in the starting operation the only large cock which is to be turned is that at the chimney blow off, and the arrangement adopted obviates the necessity of any shut off cocks between fan and ashpit and vapouriser and ashpit as are necessary in most other plants.

Another point which has been carefully arranged is that when the vapouriser is removed the depth of the generator shell is reduced to a minimum. This enables the firebrick lining to be built up quite easily, and the bricklayer is able to place each brick without difficulty.

This important point is calculated to ensure a thoroughly good job being made of all the joints of the brickwork.

A large number of these plants made in sizes up to 300 B.H.P. are successfully at work, and are found to give every satisfaction.

Regeneration from exhaust gases. A French engineer, M. J. A. Lencauchez of Paris, has constructed plants in which the waste heat of the exhaust gases from the engine are passed into a regenerator or superheater working in connection with the gas plant, and in which the air and steam passing to the furnace of the producer are superheated. The arrangement is shewn in Fig. 27 in which *e* is the exhaust pipe, *s* the exhaust heat regenerator, *g* the generator of the producer, *r* and *l* are coolers for the gas. An arrangement of this sort is calculated to secure a high efficiency but it is doubtful whether the extra complications are advisable, and in the many cases where the engine has to be a considerable distance from the plant, it could not be employed with any advantage.

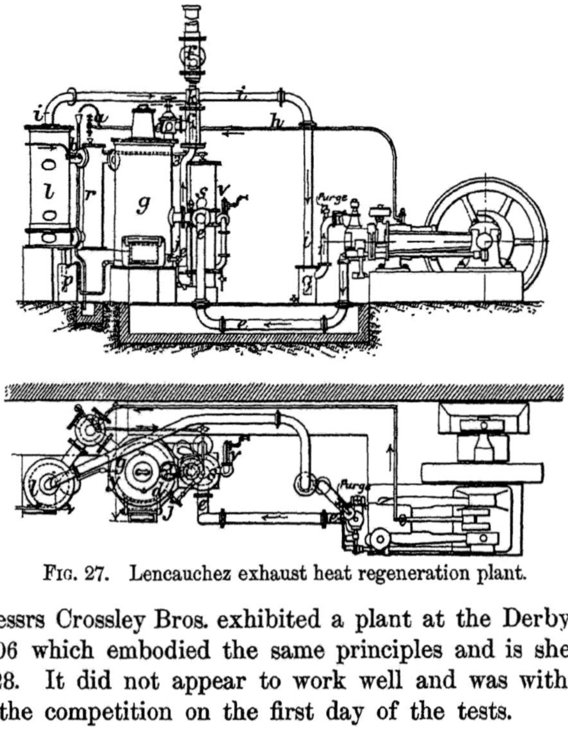

FIG. 27. Lencauchez exhaust heat regeneration plant.

Messrs Crossley Bros. exhibited a plant at the Derby trials in 1906 which embodied the same principles and is shewn in Fig. 28. It did not appear to work well and was withdrawn from the competition on the first day of the tests.

OTHER PRODUCERS.

The Hornsby, Stockport Plant. These are made by the well-known Stockport branch of Messrs R. Hornsby & Co. of Grantham. They are of the general type similar to those made by Messrs Campbell and others with top boiler and wrought iron body to the producer.

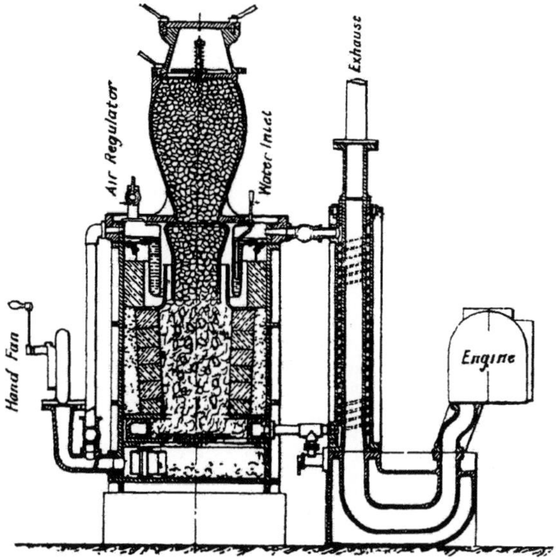

FIG. 28. Crossley exhaust heat regeneration plant.

The Mersey Engine Works Co. This plant is shewn in Fig. 29. A wrought iron jacket surrounds the body of the

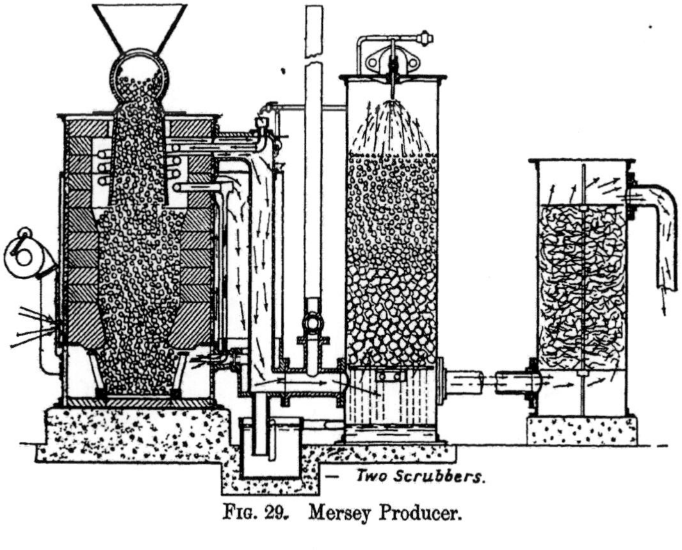

FIG. 29. Mersey Producer.

generator, and this together with external preheater, through which the hot gases pass to the scrubber, raises the temperature of the air considerably before it passes to the fire. The water is vapourised in the pipe coiled round the bell. The grate is in the form of a "devil" placed on a slightly dished iron plate to hold a little water for starting purposes.

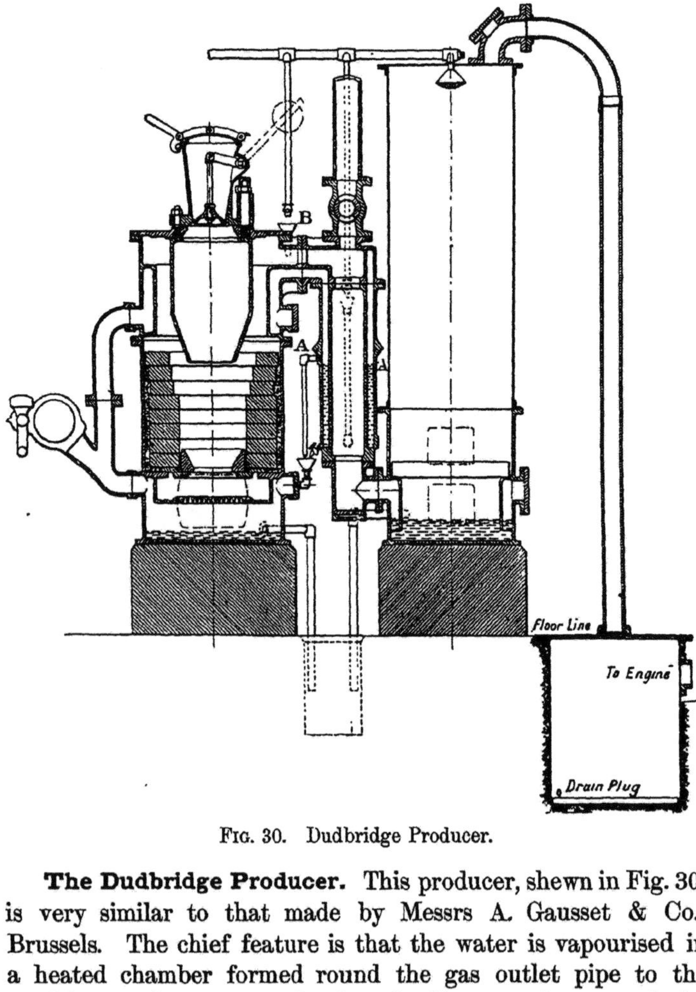

FIG. 30. Dudbridge Producer.

The Dudbridge Producer. This producer, shewn in Fig. 30, is very similar to that made by Messrs A. Gausset & Co., Brussels. The chief feature is that the water is vapourised in a heated chamber formed round the gas outlet pipe to the scrubber, and consequently the chamber at the top of the generator which usually forms the vapouriser is used as a superheater for the air and steam passing to the furnace. The other points of construction will be understood from the figure.

Suction plants are also made by several other firms, but the principles of design and construction which we have already noticed, cover the general practice adopted.

CHAPTER 5.

DESIGN AND CONSTRUCTION OF GAS COOLING AND CLEANING APPARATUS.

THE scrubbers of suction gas plants are usually of the simplest possible construction and, when of the customary form described in Chapter 1, p. 9, they merely consist of a wrought steel vertical cylinder filled with damp coke. Experience proves, however, that there are important considerations which must be carefully borne in mind if the quantity of cooling water used is to be reduced to a minimum, which is the desideratum where water is scarce and a proper cleaning effect is to be secured. The details of the scrubber connections also require careful arrangement if successful results are to be obtained under all conditions, and we have come across numerous instances where the absence of due care in the design and arrangement of such details has completely vitiated the proper working of plants which otherwise had no inherent defect.

Function of scrubbers. It is to be remembered that though a large proportion of the heat liberated in the generator, and carried off by the outgoing gases produced, is utilised in the preheating of the air passing to the furnace and the vapourisation of the water for the steam supply, this regenerative principle cannot, without undue cost, be carried far enough to extract all the heat from the gas leaving the generator, and it is consequently still at a temperature of 600° F. or over when it enters the scrubber. It is essential, however, that the gas shall be quite cool before it enters the engine, as otherwise the weight of the explosive mixture drawn into the working cylinder at each suction stroke is reduced and the power of the impulse is diminished: this of course means that if heated gas is supplied to the engine the power of the latter is seriously reduced. Hence the importance of cooling the gas thoroughly before passing it forward to the engine. In addition to being cooled the gas also requires cleaning, for there is always a certain quantity of dust and tarry matter present in it as it leaves the generator, and this must be removed before the gas is used. From these considerations it will be understood that the so-called scrubber is firstly an efficient cooler, and secondly a cleansing apparatus.

Minimum cooling water requirements. Dealing firstly with the scrubbing vessel as a cooling apparatus, the heat in the gas passing through the scrubber is taken up by the stream of water which is continually run through, usually and properly in the reverse direction to the flow of the gas, while the producer is at work, and in addition a small proportion of heat is lost by radiation from the sides of the scrubber. With the usual proportions for scrubbers at present adopted by the best makers the quantity of water required for cooling can be got down to about one gallon per B.H.P. per hour, though if the plant is being run continuously at full load $1\frac{1}{4}$ gallons is about the lowest consumption which should be counted upon. That this amount of water is actually required for cooling purposes and is a minimum can be readily shewn as follows:

Temperature of gas passing to scrubber = 650° F.
" " " from " = 65° F.
Water inlet to scrubber = 60° F.
Water outlet from scrubber = 130° F.
Weight of gas = 0·0803 lbs. per cubic ft.
Specific heat at constant pressure = 0·235
Consumption of gas per B.H.P. per hour = 65 cubic ft.
Weight " " " = 65 × 0·0803
= 5·22 lbs.
Drop in temperature of gas = (650° − 65°) = 585° F.
Rise " " " water = (130° − 60°) = 70° F.
Heat taken from gas per B.H.P. per hour
= 5·22 × 585 × 0·235 = 719 B.Th.U.

Quantity of water required to absorb this = $\frac{719}{70}$ = 10·2 lbs.

As there are 10 lbs. of water to the gallon, this comes out just over 1 gallon per B.H.P. per hour for cooling the gas, and agrees closely with practice.

If the temperature of the outlet water is allowed to rise much above 130° F. a great amount of steam tends to be made in the lower part of the scrubber due to the reduced vapour tension following on the partial vacuum in the scrubber. This vapour carries the tarry matter and dirt in the gas forward, and has to be condensed further up the scrubber so that the effective length of the latter as a cleaning apparatus is therefore considerably diminished.

Cleaning requirements. From the point of view of obtaining clean gas proper cooling is also essential. Attention has also been drawn to the rapidly increasing quantity of water vapour which can be carried by a gas as its temperature increases (see Fig. 13, Chap. 3), and naturally as the gas on its way through the scrubber is in contact with moist coke all the time it has the fullest opportunity of becoming fully saturated up to the limit of its temperature. Now each atom of water vapour can be a carrier for little globules of tar and dust, and hence the more vapour (which depends on the temperature of the outgoing gas) the greater is the tendency for the cleaning of the gas to be less efficient than it otherwise would be if the gas left the scrubber quite cool. The importance, therefore, of arranging scrubbers of sufficient capacity to admit of thorough cooling without the necessity of excessive water consumption need scarcely be further emphasised. For good results from the latter point of view, the best cooling effect is obtained with scrubbers of a good length in proportion to their diameter. For a 50 B.H.P. plant, for instance, the diameter of the scrubber would be about 2′ 6″ with a height of from 8 to 10 feet. The capacity of the scrubbers should not be much less than 0·75 *cubic feet per* B.H.P., and there is no disadvantage beyond increased first cost in their being larger than this, especially where water is scarce and the fuel used is of inferior quality.

As an apparatus for cleaning the gas, the ordinary scrubber may be considered chiefly as a dust catcher where anthracite is used, for the quantity of tarry matter present in the gas is very small indeed, and it is this fact which leads us to point out that the chief function of the scrubber is to cool the gas with this class of fuel. There is, however, a growing tendency to use cheaper fuels than Welsh anthracite with this type of gas plant, and there is no doubt that their usefulness is greatly extended by arranging them to work on gas coke, Scotch, French and Spanish anthracite. All these latter fuels, however, give off a considerably greater percentage of volatile matter, chiefly in the form of tarry compounds, which, for engine work, must be cleaned out of the gas: otherwise considerable trouble is caused by the inlet valves of the engine sticking, especially if it is stopped and allowed to get cool without the valve spindles being cleaned from any tar deposit which may have formed upon them during the run. As a cleaning material the ordinary coke, with which it is customary to pack the scrubbers, is not in itself sufficient for the dirtier fuels referred to above for plants over 30 B.H.P., and in view of the recent considerable advance in the price of Welsh anthracite it seems almost certain that there will be considerable developments during the next few years in the improvement of the scrubbers as a cleaning apparatus. Hitherto, with the common use of Welsh anthracite, very little cleaning has been required and little thought has accordingly been given to the design of the scrubbers, but it is in their improvement to make the use of cheaper fuels more easily possible that further attention is now being given.

Messrs Pierson of Paris have adopted the type of scrubber introduced by M. F. Chevalet of Troyes, for their suction plants which are claimed to work successfully on the "lean" French anthracites which contain a considerable percentage of tarry matter. These scrubbers are of cast iron, built up in sections of from 18 to 24 inches deep, which are put together by spigot and socket joints. Between each section there is a flat perforated plate for distributing the water and breaking up the gas. Stiff spiral shavings of wood are used to pack these scrubbers instead of coke, but it is not apparent why, under equal conditions, the former material should be better or cheaper than the latter. Side doors are provided all the way up, and these admit of the easy removal of the wood in each section without disturbing that in the adjoining compartments.

General construction. Scrubbers may conveniently be made of light mild steel plates. The rivets should be closely pitched and every care taken to ensure absolute tightness throughout. The doors and facings provided for pipes and other connections are usually of cast iron, and in the best plants they are machined on the joints. Some amount of corrosion is to be expected in the lower part of the scrubber where the water lies about, and it is good practice to coat the interiors with Dr Angus Smith's or other suitable anti-corrosive paint. When being erected proper care should be exercised to ensure that the scrubber stands quite vertically; otherwise the cooling water when dropping from the distributor tends to fall all to one side, and under these conditions the other side gets very hot, and the scrubber generally is only half effective.

Sawdust scrubbers. Sawdust is a good material for removing the final traces of tar, and the use of sawdust scrubbers is now almost general. In designing sawdust scrubbers it is well to remember that the best results are obtained by distributing the gas over a comparatively large area or layer of sawdust through which it passes at a low velocity. In this way a small bulk of sawdust, easily renewable, can be made very effective. Usual forms of Dowson sawdust scrubbers are given in Fig. 31. Messrs Crossley adopt, in their smaller plants, the plan of having sawdust in the upper part of their coke scrubber, but we doubt whether the reduction of the capacity of the latter as a cooler is altogether wise.

Tar extractors. It is shewn more fully in Chapter 15, p. 223, that a considerable proportion of tar can be precipitated out of the gas by breaking it up into fine streams and causing these to impinge against solid plates in series, the direction of motion being sharply changed between each plate. Fig. 32 below shews the National combined impact tar extractor and sawdust scrubber designed on the foregoing principles, and which is found to work well. The dash plates are arranged to be easily removable for cleaning purposes, and special attention has been given to ensure that the gas passes evenly through the sawdust for the final cleansing.

Fig. 33 shews Messrs Tangye's tar extractor for the same purpose. The internal tar catching surfaces can be conveniently withdrawn for cleaning purposes by removing the top cover.

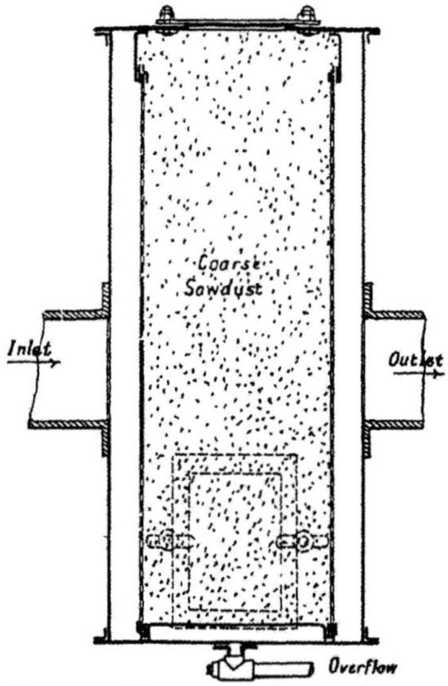

FIG. 31. Dowson sawdust scrubbers.

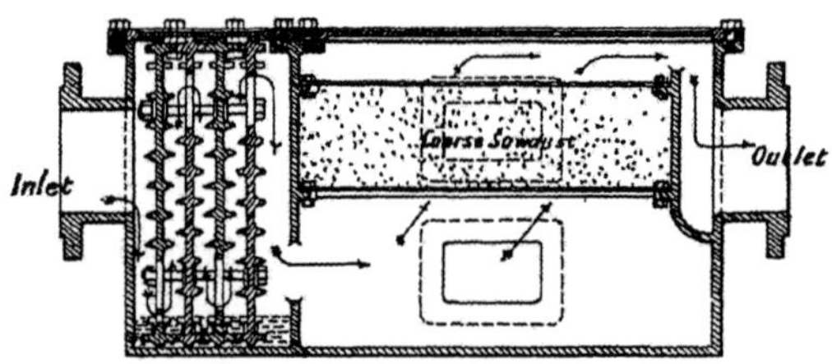

FIG. 32. National tar extractor.

It will, of course, be understood that these appliances are unnecessary in plants up to 100 B.H.P. excepting where the fuel used is gas coke or poor anthracite. They are now fitted, however, to all plants of 150 B.H.P. and over.

Centrifugal tar extractors have also been applied by Crossley Bros. and other makers, but users prefer to have static extractors if they are sufficiently effective.

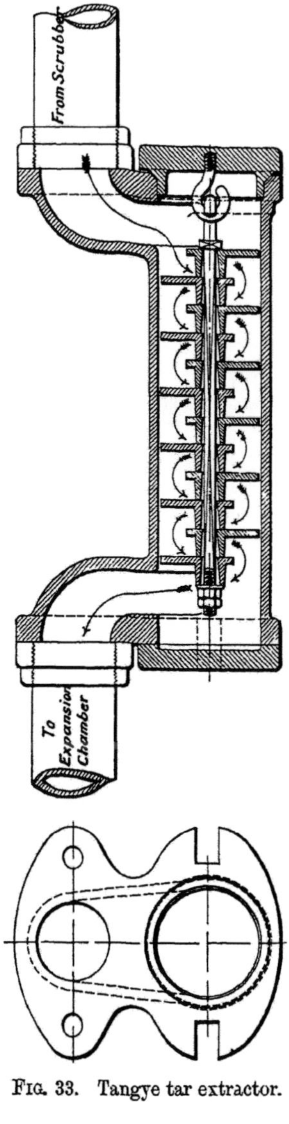

FIG. 33. Tangye tar extractor.

Water connections. The methods by which the cooling water is introduced, and the outflow arranged, require careful consideration, for, though simple in themselves, they must be properly carried out.

It has been customary to fit a star-shaped wrought iron distributor at the top of the scrubber, to the centre of which the water supply is brought as in Fig. 34. Each radial arm of the star is perforated with small holes through which the water is supposed to spurt. The idea of spurting, however, is totally

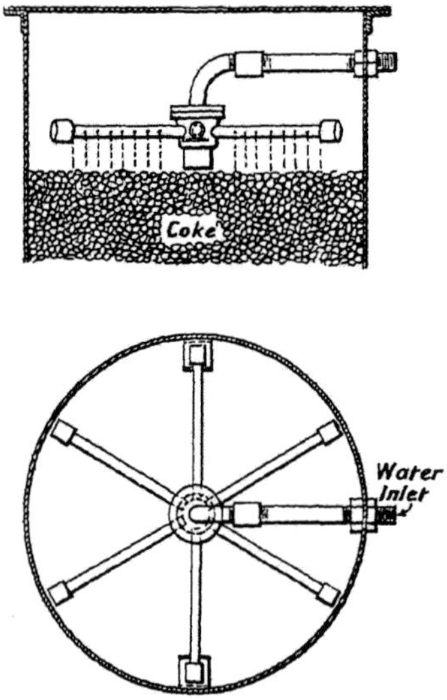

FIG. 34. Usual type water distributor at top of scrubber.

erroneous, for when the water supply is cut down to the required rate of about one gallon per B.H.P. per hour, the water gently dribbles out and collects in beads, which, dropping promiscuously, make a very uncertain distribution over the full cross-sectional area of the scrubber. After a while the small holes become rusted up, and shortly afterwards the sprinkler ceases to work altogether.

In the National patent plants this difficulty has been got over by distributing the water from an open cast iron sprinkling dish into which the supply is brought at two places through a good sized pipe. As the dish fills up the water flows at notches arranged immediately over dripping points which effectively distribute the water equally over the whole area of the scrubber. There are no small orifices to make up, and this simple appliance which is shewn in Fig. 35 works admirably.

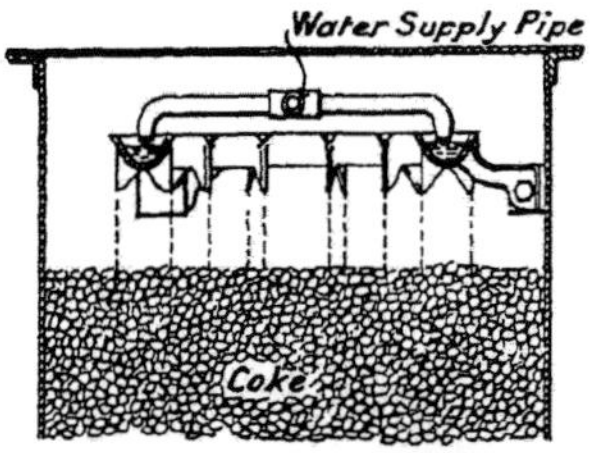

FIG. 35. National type water distributor.

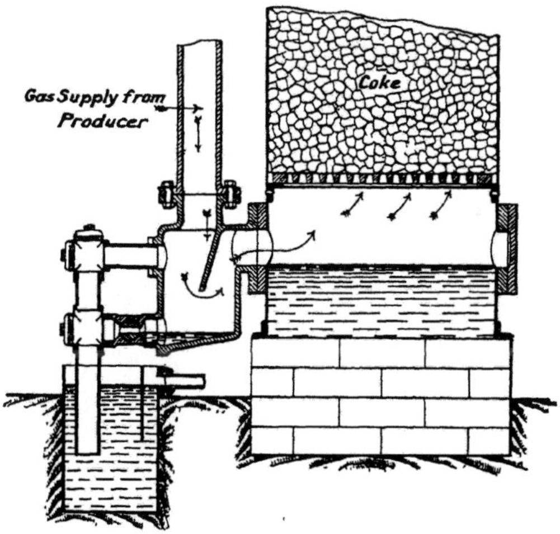

FIG. 36. Detail of gas connection to scrubber (Tangye).

Introduction of the gas to the scrubber. Typical designs of inlet connections, whereby the outlet pipe containing the hot and dirty gas from the producer is brought into the scrubber, are shewn in Figs. 36, 37 and 38. In the National plants there is a shallow water seal through which the gas is forced before passing into the scrubber. This has a double advantage, for it

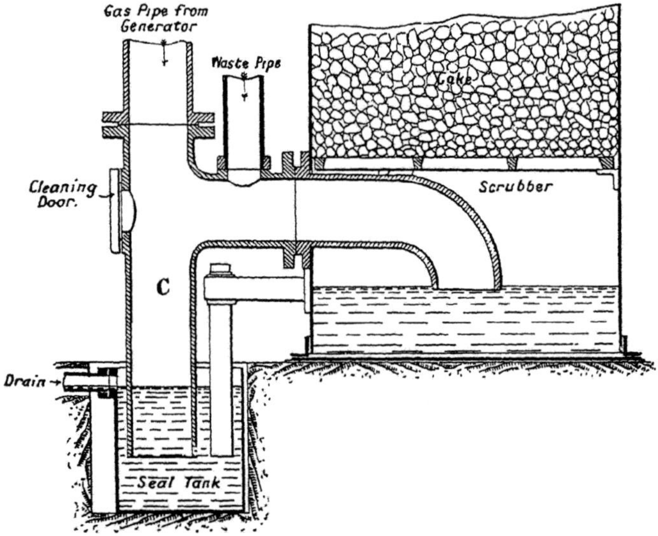

FIG. 37. Detail of gas connection to scrubber: internal seal (National).

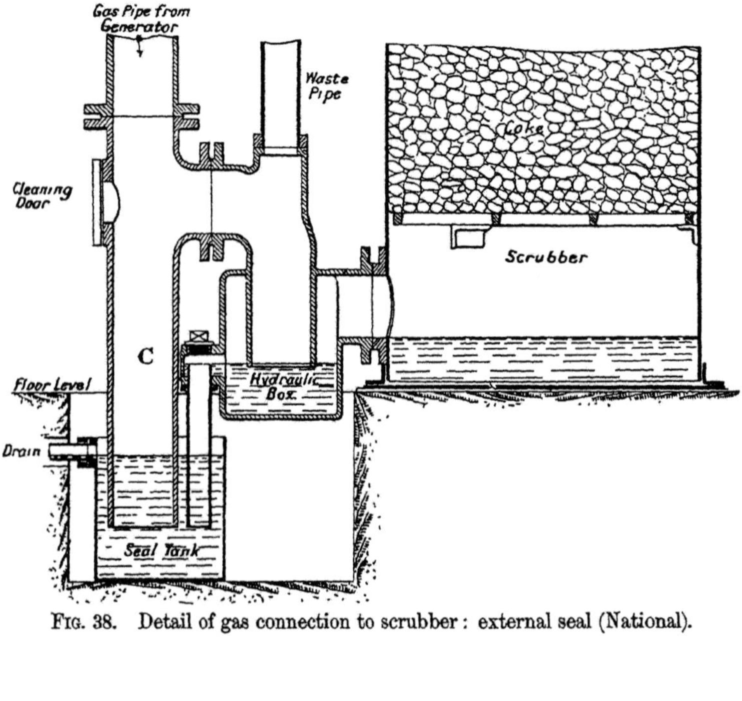

FIG. 38. Detail of gas connection to scrubber: external seal (National).

ensures that a large portion of the coal dust in the gas carried over from the generator is precipitated direct into the water at the bottom of the scrubber before passing through the coke or other cleaning material, and this tends to the continued efficiency of the latter over several months, whereas otherwise it would rapidly become clogged up. In addition, the seal prevents the gas contained in the scrubber, after the plant is stopped, from leaking back into the generator whilst the latter is being cleaned out after each week's run. Otherwise the attendant working in or about the generator might be injuriously affected by this gas which is of a very poisonous nature. In other plants the water seal appears to be omitted, but, for the reasons given, the advisability of this is open to serious doubt.

The depth of the seal need only be from $\frac{1}{4}''$ to $\frac{1}{2}''$ and the height is regulated by the level of the drain-pipe provided with a screwed connection for the purpose. The method of arranging the gas inlet pipe with an open end C (Figs. 37, 38) dipping into a seal pot is the best and safest, for in the event of a slight explosion in the top of the generator, with which this pipe is in free communication, due to air being allowed to leak in through inadvertence on the part of the attendant, any pressure set up is easily relieved through the open end of the pipe without damage to any of the pipes or fittings, and the water is merely blown out of the seal pot.

Cleaning doors. Convenient cleaning doors should be fitted to all gas pipes about the generator and scrubber, so that accumulations of dust or tar can be easily removed. In all scrubbers of 50 H.P. and over it is advisable to fit two supporting trays for carrying the coke with a filling and draw-off door respectively to each division of the scrubber thereby formed. The coke in the lower portion of the scrubber naturally becomes clogged up more rapidly than that above, and it is usual to replace the former about every five or six weeks and the latter once in six to twelve months according to the character of the fuel used.

Disposal of the effluent water from scrubbers. In general, as the effluent water from suction plants contains very little tarry matter or dirt, it is turned into the ordinary town drains, and causes no nuisance. It sometimes happens, however, that due to the chemical composition of the fuel which varies considerably as supplied commercially, an unpleasant smell arises from these effluents, and this odour, finding its way out of the sewer vents, becomes a matter of complaint. In many cases this smell is due principally to the presence of free ammonia in the water which is gradually given off as the effluent passes away. The proportion of ammonia may be as much as 33 parts in 100000. This ammonia, with certain other contained compounds which tend to cause a nasty unpleasant smell, may be driven off by aerating and heating the effluent in a flue which will carry away the fumes.

On a practical scale this is done by spraying the effluent water over the exhaust box of the engine which for this purpose is placed in a suitable flue up which a current of air is induced by the heat of the box. The fumes escape up the flue after which the effluent, which is then quite unobjectionable, may pass to the drain. In this simple way the required aerating and heating of objectionable effluents are carried out at a cost of a few extra feet of piping.

Any acid present in effluent water can be neutralised by placing a few lumps of chalk ($CaCO_3$) in the bottom of the scrubbers, and thus by neutralising the acid and driving out the fumes in the manner suggested above, a storage of cooling water for the scrubbers can be made so as to use the water over and over again, only making up with fresh water the proportion which is evaporated in the deodorizing process.

Messrs Tangye have worked out these principles in the manner shewn in Fig. 39. Two tanks are provided, one open and one closed. The water, after passing from the scrubber to the drain box, is gravitated through pipe (1) to the closed tank, where it is filtered so as to cleanse it from dust and any other solid matter it may contain; the filtered water is then taken through pipe (2) by a small pump and forced along pipe (3) through the cooling jacket of the cylinder, after which it is passed by pipe (4) to a combined exhaust chamber and deodorizer, in which the water is freed from the objectionable fumes by the heat of the exhaust chamber, and the gases are passed to the atmosphere with the exhaust of the engine cylinder. The purified water then passes through pipe (5) to the open tank, from which it is pumped through pipes (6) and (7) to the scrubber for cooling the gas as it is made in the generator, passing to the seal box to be afterwards purified as previously described.

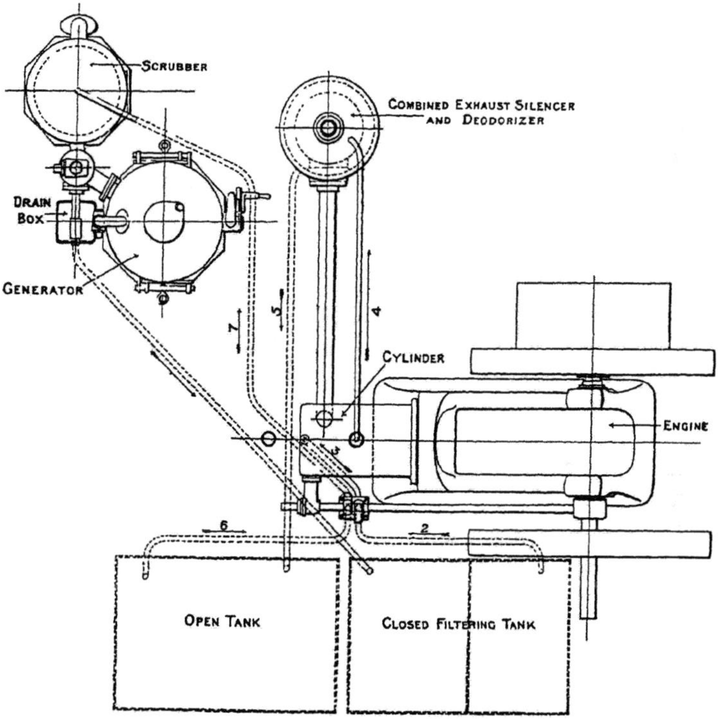

FIG. 39. Diagram shewing arrangement of "Deodorizer."

The arrangement has been proved successful; and, in localities where water is scarce, the slightly additional cost will soon be repaid.

In general it may be stated that, when Gas Works coke is used as fuel in the generator, the effluent will shew distinct traces of acid, and with this fuel we advise the introduction of chalk into the bottom of the wrought iron scrubbers to neutralise same. This acid is not usually present when anthracite is used.

It sometimes happens that sulphuretted hydrogen appears in the effluents, and these are in such cases passed over a filtering apparatus filled with alternate layers consisting of a mixture of oxide of iron (the same as used in Gas Works for gas purification) and wood shavings. The object of the wood shavings is to prevent the oxide clogging together, in which event the water would not trickle through, and it is usual to have a layer of pebbles at the top and bottom of the filter to prevent the oxide mixture being washed away. The oxide should be turned out every three or four months to be revivified by being exposed to the air, a fresh lot being put in the filter meanwhile. Two or three weeks are required for revivification.

Effect of gas plant effluents on sewage beds. There would appear to be nothing in the effluent water from these suction plants which when diluted with other general sewage is sufficient to interfere with the proper action of the bacteriological sewage beds.

CHAPTER 6.

OTHER IMPORTANT DETAILS.

Water supply to generator. In the preceding chapters such frequent reference has been made to the vital importance of a proper steam supply for mixing with the ingoing air passing to the furnace of the generator for the double purpose of keeping down the fire and of enriching the gas, that the necessity of making careful arrangements in connection with the water feed from which this steam is produced will be fully appreciated.

In considering this matter, it is well in the first place to realise practically what the required amount of water actually means. This latter is 0·75 lbs. per B.H.P. per hour, so that in a plant developing 100 B.H.P. the total amount of water required for the generator is 75 lbs. or 7½ gallons. Now a uniform discharge from a regulating valve at the rate of 7½ gallons per hour is little more than a trickle, and it is really the fineness of the discharge which makes it necessary to ensure that, under fluctuation of pressure in the water supply from which this feed is drawn, the discharge will not cease altogether.

In those plants fitted with vapourisers arranged to carry a volume of water as in an ordinary boiler, an overflow is always arranged so as to maintain the proper water level, and this means that considerably more water is allowed to pass in as feed than is actually required for the steam supply, but even if twice as much water is used in this way as is required for steam, the discharge is still very small, and if connected direct to the town's mains, wherein the pressure is invariably in a state of constant fluctuation, trouble may be expected.

The same considerations apply to the water supply to the scrubbers and no intelligent user likes to feel that it is necessary to waste on the average twice as much water as is actually required simply to ensure a sufficiency at odd periods of the day. With the object, therefore, of ensuring a constant and regular supply of water to the gas plant, the town's supply should be brought to a small overhead tank fitted with a ball-cock, from which the feed pipes to generator and scrubbers are taken. Fig. 40 shews a usual arrangement of water connections which will be found to give the desired result. When the water supply is drawn from a reservoir or river, a larger overhead tank is used, and this should be of sufficient capacity to allow of a slight feed to the vapouriser being kept up over a week-end while the plant may be shut down but the fire kept in. At such times the pump drawing from the source of supply and discharging into the overhead feed tank will very likely be stopped, but the vapouriser must have some water supply all the time the fire is going. Otherwise as soon as the reserve of water is evaporated which is left in the vapouriser when the plant stops working, heat cracks may develope, especially if the water supply is subsequently suddenly applied. Hence the advisability of having a tank sufficiently large to allow of a slight dribble being run into the vapouriser overnight or during week-ends.

The water supply to the generator should always be arranged as an open jet discharging into a funnel, so as to be under easy observation. This not only allows of an approximately correct adjustment of the quantity passing in, but any stoppage of the water supply is more likely to be noticed before any serious consequences can arise, such as a stoppage of the engine through the poorer gas made, or the burning out of the fire-bars through the excessive heat of the fire when steam is not mixed with the incoming air.

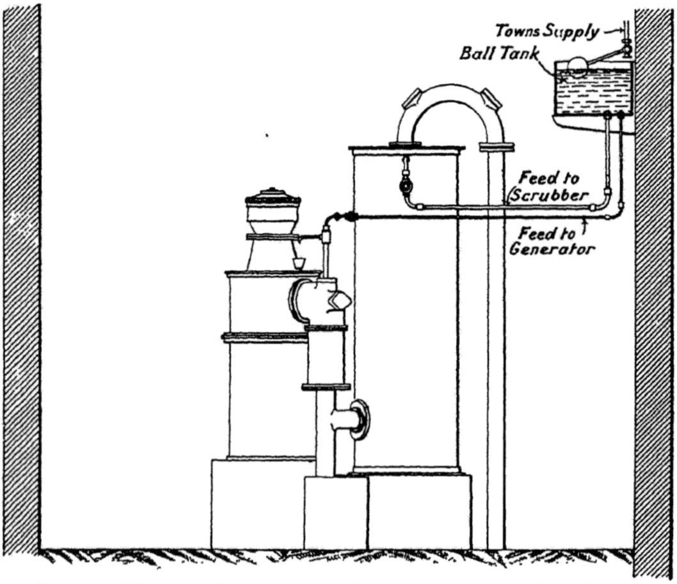

FIG. 40. Diagram of water connections to generator and scrubber.

Before leaving this question of the water supply to the generator, it might be pointed out that where vapourisers of the flash boiler type are employed and the water feed is drawn from a tank fitted with a ball-cock, the presence of the latter ensures a constant and regular rate of supply. The feed may therefore be cut down to the exact requirements for making the richest gas—neither more nor less, and the steam supply is therefore under proper control to secure the best results.

Automatic regulation of water feed. In the latter part of Chapter 4 we have already dealt with the advantages and disadvantages of automatic regulation of the water feed to the generator, and though, for the reasons therein given, we do not consider that automatic regulation confers any gain whatever, it is interesting to observe how it has been carried out in practice. One of the simplest and most ingenious arrangements is that embodied in the plant made by Messrs Fielding & Platt of Gloucester. Figs. 41 and 42 are taken from the

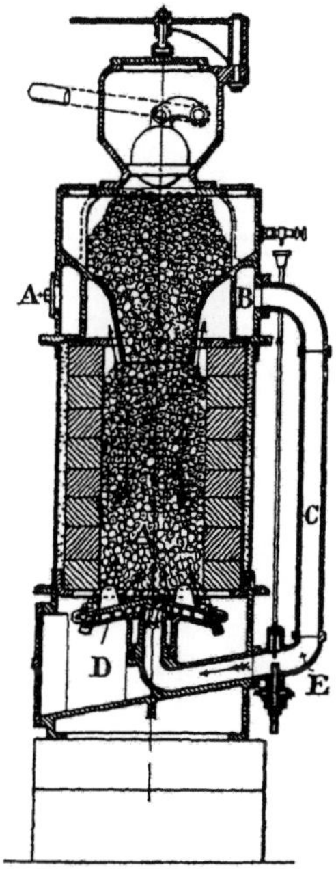

FIG. 41. Fielding and Platt generator.

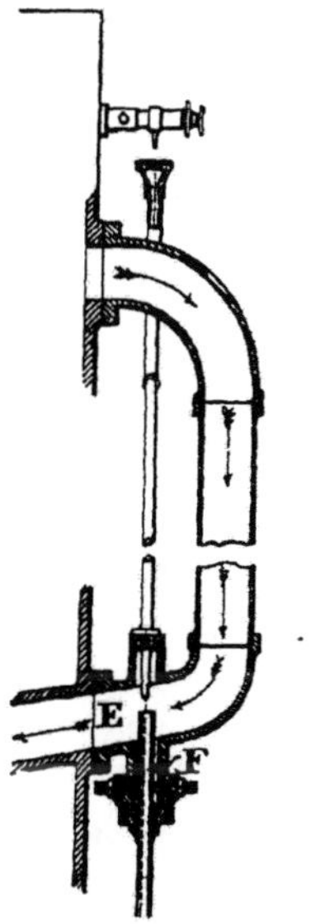

FIG. 42. Fielding and Platt automatic water regulator to generator.

published descriptions of the plant exhibited by this firm at the Derby Trials. The former gives a section through the generator, from which it will be seen that the upper part of the latter, which usually forms the vapouriser in other plants, is here used simply as an air heater. Air enters at the opening A and passes round the heated chamber B and thence down the pipe C on its way to the fire. At E the water supply in a fine stream drops into the waste pipe F, excepting at each rush of air through the pipe C corresponding to each suction stroke of the engine, when the impact of this air is sufficient to divert the water aside so that it misses the overflow and passes forward towards the fire instead. Complete vapourisation and some superheating is supposed to be effected by contact of the air and water with the ribbed underside of the cast iron plate supporting the fire. It is extremely doubtful, however, whether the steam supply with this arrangement is anything like adequate. In the first place the water has to issue from the nozzle at E in an exceedingly fine stream in order that the required action may be ensured, and while in plants of different size the velocity of air through this pipe should be kept practically the same, a much larger stream of water will be required for the larger plants than for the smaller ones. A heavier stream will not, however, be carried forward with the same facility as a lighter one, and from this point of view the system would appear to be somewhat precarious. Apart from this altogether there would seem to be some uncertainty about the complete vapourisation of the water, even after it is carried forward by the air, as the heating surface provided is considerably less than other makers have found to be necessary, and the results of the Derby Trials give some support to this view, for the clinker found in this plant is described as being "hard."

In the French plant patented by J. Delassue the pressure of the exhaust of the engine is made to operate a valve or cock which controls the supply of steam to the fire. As each exhaust stroke corresponds to a suction stroke of the engine, this means that the amount of steam allowed to pass to the fire is in exact proportion to the load. Even if there were any advantage in such an arrangement, the very considerable distance at which the gas plant has frequently to be placed from the engine will introduce factors which will tend to cause trouble. A section through this plant is shewn in Fig. 43.

Firebars and grates. In many industries it is necessary to keep the machinery in operation for the whole working week, night and day, without stoppage. This means a non-stop run of at least 144 hours. When suction gas plants of over 80 B.H.P. are called upon to meet these conditions they should be amply large for the specified duty, and in addition some simple means of keeping the grate clear of ash and clinker must be provided.

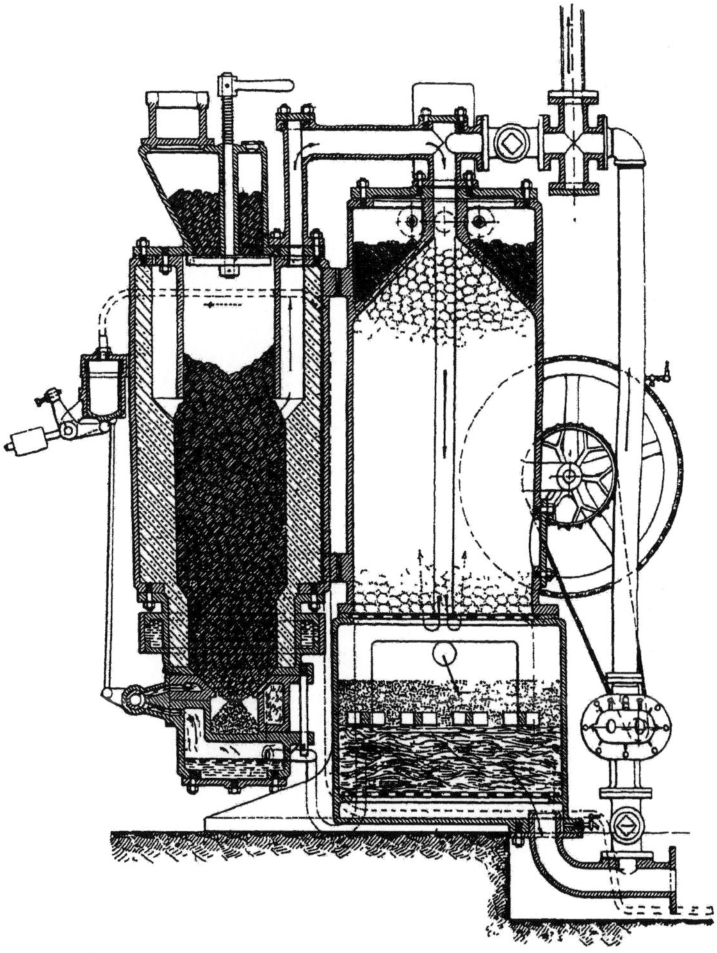

FIG. 43. Delassue plant with automatic water control worked by exhaust pressure.

Revolving grates, which, when slowly turned, cause the ash, etc. to fall away, are used by some makers, and a customary design is that shewn in Fig. 44.

In the Dowson plant a form of shaking grate is used consisting of special firebars which can be moved horizontally through a small distance by working external levers and handles. This arrangement is shewn in Fig. 45. Against this form of grate, however, it is urged, with some truth, that as the resistance, due to the weight of the whole of the fuel, is opposed to the motion of the bars they are apt to stick, as even under good conditions they are not easily moved.

This difficulty has been overcome in a very simple manner in the patent firegrates of the National plants shewn in Fig. 46.

The grate is arranged with fixed and rotating firebars placed alternately. The latter are triangular in section and have pro-

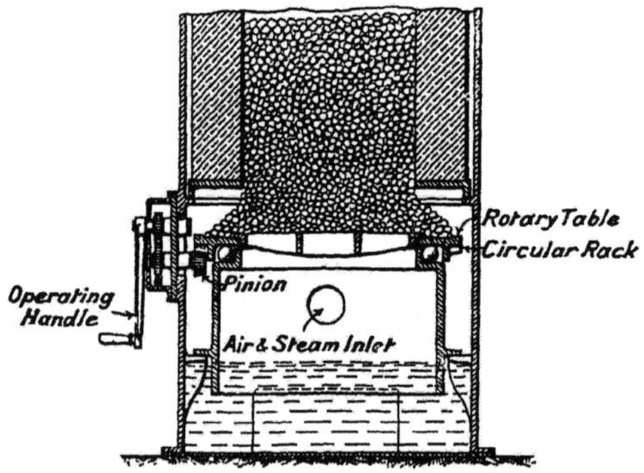

FIG. 44. Diagram of a revolving grate.

jecting ends brought through the dead plate. A handle can be applied to these ends whenever it is necessary to shake down the ashes and as only one bar is operated on at one time, there is no difficulty in effecting the necessary movement, and moreover the fire is cleaned so gradually that the gas produced is kept of good and uniform quality. There is a secondary fire door in the dead plate for the usual purpose of cleaning, and air is prevented from leaking in past the projecting ends of the bars by a main door, which makes an air tight joint all round similar to the ordinary fire door.

Fire doors and ashpit doors. It has now become recognised that these should be a thoroughly good job. The ashpit door should be quite separate from the fire door and both doors should be carried on substantial hinges so that they can be easily opened. The faces of the doors should be machined so that no luting or asbestos is required to make an air tight joint when the door is shut and secured.

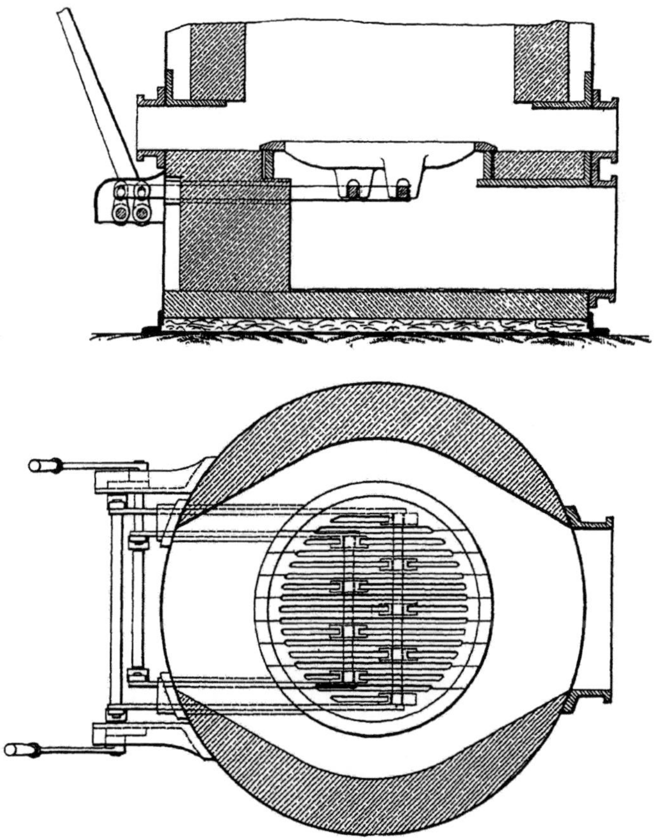

FIG. 45. Dowson fire-grate with moving firebars.

In generators of 75 H.P. and upwards there should be two fire doors opposite each other to assist the thorough cleaning out of the furnace from time to time.

Hand blowers for starting purposes. This is an important detail in so far as the starting up of the gas plant depends upon it. It usually takes anything from 10 to 30 minutes to get a burnable gas after lighting the fire or after the plant has been shut down overnight and the fatigue of having to turn a faulty or inefficient hand blower over half an

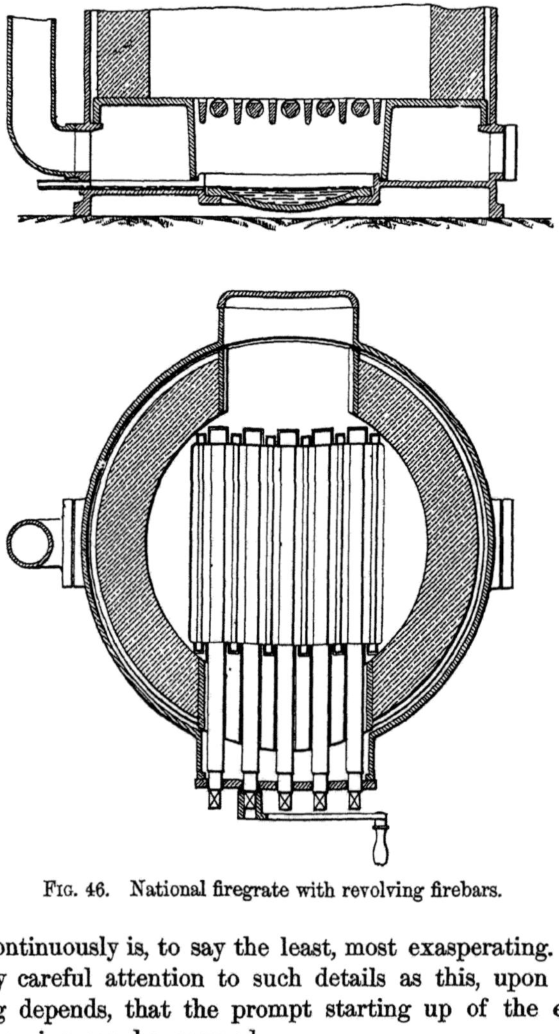

FIG. 46. National firegrate with revolving firebars.

hour continuously is, to say the least, most exasperating. It is only by careful attention to such details as this, upon which starting depends, that the prompt starting up of the engine each morning can be ensured.

The usual hand-geared fan is a very poor apparatus and this caused the author to design a special fan for the National plants

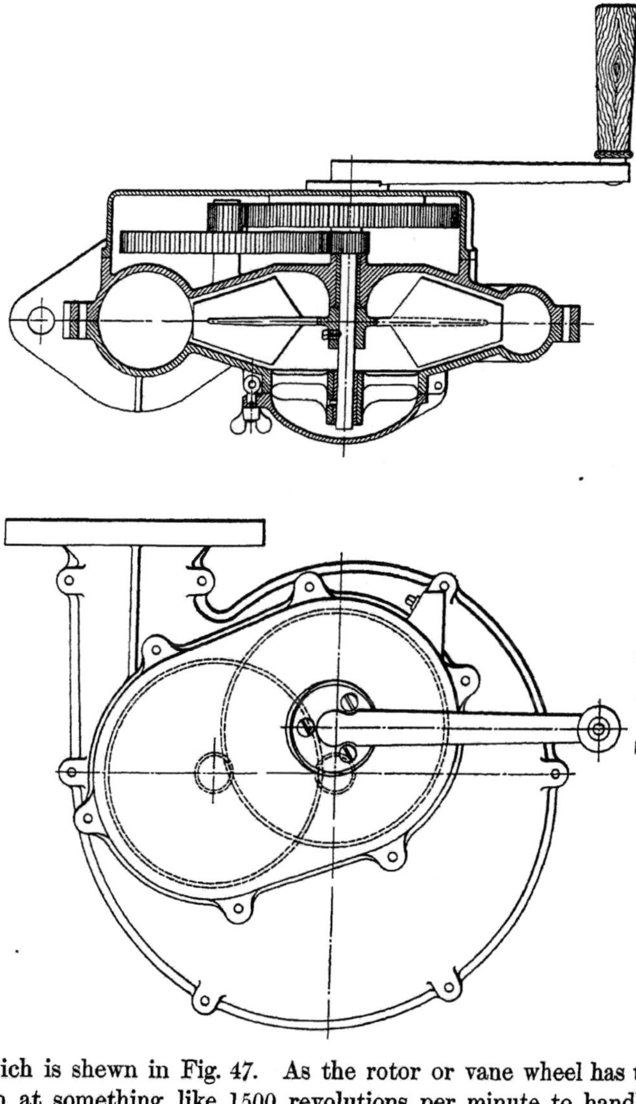

Fig. 47. Geared hand blower for starting purposes.

which is shewn in Fig. 47. As the rotor or vane wheel has to run at something like 1500 revolutions per minute to handle the required volume of air without having a large and clumsy apparatus, it will be realised that the gearing to speed up the revolutions from about 40 per minute at the handle to the rotor speed just named, requires to be well made and carefully designed to reduce friction. In the fan illustrated in Fig. 47 there are two sets of gear wheels with machine cut teeth carefully shaped and working in an oil bath. Special provision is made to prevent the oil leaking from the gear case either to the rotor or to the outside of the cover. The rotor must be made exceedingly light so as to punish the fine gear wheel teeth as little as possible while it is being accelerated at starting. The pinions in gear trains should be of the best mild steel. A fan with a 3″ outlet and other parts in proportion is about the maximum size an ordinary man can turn continuously over the usual fifteen minutes required for blowing the fire into proper condition for starting. Smaller fans may be used on plants below 50 B.H.P.

Power blowers. These will be found a great convenience for starting plants of 100 B.H.P. and upwards. They usually consist of a direct connected electrically driven fan with speed regulating switch to the motor. In arranging for them, a capacity equal to a discharge of not less than 2 cubic feet per minute per B.H.P. against 3 inches of water gauge pressure should be allowed. An alternative method of driving such a power blower is by the auxiliary gas or petrol engine which is frequently fitted for working the air compressor in connection with the starting apparatus of the main gas engine. Another alternative to either of the two named for obtaining power for operating starting fans, is to use a small Pelton wheel driven from the general water supply. This method has the merit of being both cheap and effective.

Expansion boxes. These are simply hollow receivers inserted in the gas main near the engine. Their object is to provide a good volume of gas close at hand which will easily follow up the suction action of the piston so that the minimum vacuum effect produced by the forward movement of the latter will cause the charge of gas to flow into the cylinder. In the absence of such a receiver it would be necessary for the suction action of the piston just referred to, to accelerate the column of gas in the pipes, scrubbers, etc. before the charge would be drawn into the cylinder. During the time occupied in so doing the piston would have moved through the greater portion of the suction stroke, and consequently by the time the latter was completed the cylinder would be without the proper amount of combustible mixture. With an expansion box close at hand, however, the engine draws its charge of gas readily and, during the compression power and exhaust strokes which follow, the expansion box has ample time in which to fill up again. To express the matter in another way, under a given small difference of pressure between inside of engine cylinder and gas mains, the gas will flow more readily in the available time from the expansion box close at hand than when, in the absence of the same, the inertia of the gas in mains and scrubbers has to be overcome.

Expansion boxes are usually cylindrical in form and made either in wrought iron or cast iron with inlet and outlet branches. The gas mains should be sloped towards the expansion box and a drain cock fitted to the latter so that any moisture collecting in the pipes can be readily drawn off. There is also a blow off cock and waste pipe fitted so that air or bad gas can be conveniently blown out of mains and box when starting up the plant prior to the actual starting of the engine.

The capacity of expansion boxes should not be less than one-half the volume of the working cylinder of the engine to which they are connected, and they may with advantage be larger than this where circumstances allow.

CHAPTER 7.

SUCTION PLANTS FOR BITUMINOUS FUELS.

MOST gas plant makers have taken out patents at different times for suction plants to work with bituminous fuels, but up to the present it is safe to say that their efforts have not culminated in a really commercial apparatus. The general difficulties resulting from the use of bituminous fuel in producers which depend on the suction action of the engine to regulate the make of gas have already been indicated in Chapter 1. It was pointed out there that the quantity of ash and clinker coming from this fuel is so considerable as to necessitate regular poking in the producer, and as a matter of fact this must be done at least every fifteen minutes while the plant is at work. This operation alone requires the constant attendance of a man at the producer, which it will be noted is very different from the amount of attention required by the ordinary suction producer using anthracite or coke, in which the only operation necessary is to introduce a fresh charge of fuel every two or three hours.

Another great difficulty—the chief one where the gas is to be used for engine work—lies in the effective removal of the tar from the gas produced from bituminous fuel. The light hydrocarbons in the latter readily distil off from the fresh fuel in the upper part of the producer under the action of the temperature set up there while the plant is working, and in the form of tarry vapour they mix with the outgoing gas passing away from the producer. Their presence in the gas thereafter necessitates very elaborate cleaning apparatus, which introduces heavy friction and consequent back pressure against the suction effect of the engine, and this makes it impossible to get the proper amount of gas into the working cylinder during the suction stroke.

The troubles in the generator can be overcome to a large extent by careful arrangements for poking the fire and withdrawing the ash and clinker so as to prevent the fire becoming choked. The difficulty from the tar in the gas made is, however, more troublesome, and it is on the problems of its elimination that most thought has been expended.

In this connection the principal idea tried up to the present and which has been developed in various ways, is to work the producer with a down-draught instead of an up-draught on the principle of compelling the tarry vapours to pass through a depth of incandescent fuel so that they may there be converted into fixed gases. Though a considerable proportion of the tarry vapours can be fixed in this way, it is well established now that it is impossible, under any temperature obtaining in the producer, to fix all such, and so far as any figures are available it would appear that in practice with ordinary bituminous fuel only about one-third of the tar can be reduced in this way. Hence, so far as suction type producers are concerned, the only down-draught generators which have actually been put to use for

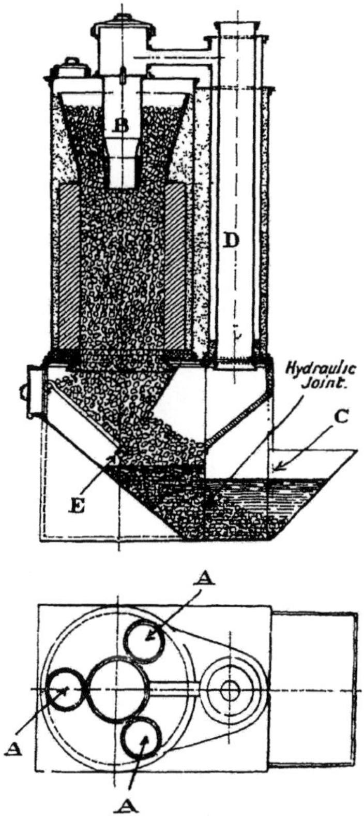

FIG. 48. Deschamps down-draught producer for bituminous fuel.

engine work have been operated with semi-bituminous fuel such as that found in Belgium and France.

The Deschamps Producer. M. Jules Deschamps has worked out a down-draught suction producer in general accordance with Fig. 48 which is claimed to work successfully with peat, lignite and bituminous fuels.

It will be seen that the fresh fuel is fed into the top of the producer through three openings A, which are normally closed with removable covers while the plant is at work, except when fresh fuel is added or poking required. The incoming air and steam are led into the centre of the top of the producer through the supply pipe B, which is carried down to about one-third of the depth of the furnace. The fresh fuel is kept at a height well above where the air enters the fire, so that the volatile hydrocarbons are distilled before the fire is reached. The ashes and clinker are arranged to fall into, and to be withdrawn from, the water lute C, and the water in the latter is made to form a hydraulic joint to prevent air being drawn along with the gas, which passes off at the bottom of the producer and up the pipe D. The latter is jacketed with a cover pipe and in the annulus formed between the two, the air passing to the furnace is preheated and the steam raised. The sloping firebars are hinged at their upper end, and can be moved up and down through a certain degree by a cam shaft E at the lower end, and in this way the ashes are shaken down from the furnace into the lute from time to time. A certain amount of the steam formed by the hot ashes falling into the lute is also caused to mix with the air passing to the fire. Experiments with lignite, coke, cinders from locomotives, and bituminous coal were made on one of these producers and a 25 H.P. engine. The trials with each class of fuel extended over about 12 days, and were in each case considered satisfactory. The chief proportion of tar was formed in getting the plant to work in the first instance: when the fire was got going with clean coke before putting on the bituminous fuel, this trouble was eliminated. The washing apparatus consisted of an ordinary coke scrubber and a small centrifugal fan worked by an electric motor. In most of the experiments the latter was used, and the gas discharged into a small holder, and any excess of gas produced over what the engine required was burned to waste. Of course the centrifugal tar extractor would have an important effect on the cleansing of the gas, but M. Deschamps states that several of the experiments were repeated with the plant operated by the suction of the engine alone, and the valves of the engine were found to require no more cleaning than is usual when anthracite is used in the ordinary suction producer. In order to secure a regular quality of gas, it is essential that the ashes and clinker be shaken down at regular intervals, otherwise the fire is disturbed too much, and the quality of the gas interfered with. In one case quoted, the coal used contained 39·9 °/₀ of volatile matter, and 23 °/₀ of fusible ash, which latter especially tends to form clinker, and becomes attached to the sides of the furnace: the trial lasted 10 consecutive days, during the last four of which there was no stoppage night or day. The engine valves were not touched until after the trial, when they were found to be clean, and no clinker was found to be adhering to the sides of the furnace of the producer.

Whilst these experiments are most encouraging, it must be realised that further trials under working conditions and extending over several months are necessary before any firm opinion on the merits of this producer can be given with confidence.

Fichet and Heurtey Producers. An extension of the down-draught principle is to combine it with an up-draught in the lower part of the same producer, and Mr Dowson has made an experimental plant on these lines. Messrs Fichet & Heurtey actually shewed at the Liege Exhibition a 400 H.P. producer designed by them to operate on this double-draught system, and it appeared to work well with semi-bituminous coal containing about 12 °/₀ of volatile matter. It is extremely doubtful, however, whether it would be sufficiently successful in dealing with the ordinary bituminous coal available in this country.

Fig. 49 represents a section through the Fichet and Heurtey producer referred to. In general arrangement it is something like two generators placed together one on top of the other and fed by the same fuel and having the same gas outlet at A, but two separate fires at B and C respectively. The combustion is normal in the lower generator—that is to say, from bottom to top—but in the upper generator it is reversed and so arranged that the hot zone of the upper fire is maintained at B. The coal container D above the upper fire may therefore be regarded as a distilling chamber for the fresh fuel, and the light hydrocarbons which are first driven off this fresh fuel are carried forward along with a certain amount of air admitted into the top of the

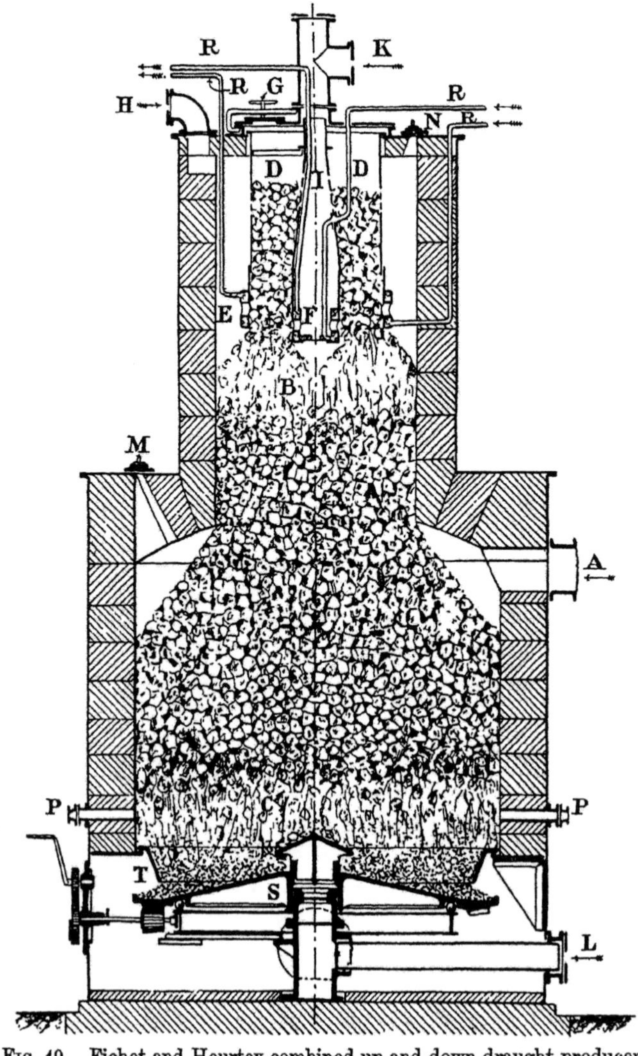

FIG. 49. Fichet and Heurtey combined up and down draught producer.

container and passed through the incandescent fuel in the upper fire before reaching the outlet A. This to a considerable extent secures their decomposition into fixed gases before they leave the generator. The fuel is only partially decomposed in the upper furnace and consequently the bottom generator is in the position of being fed with distilled coal and works like a generator, using coke or other non-bituminous coal.

Special care is taken with the arrangements for introducing the air into the upper furnace so as to keep the hot zone there in the proper position to ensure that distillation *only* takes place in the coal container, and consequently to better prevent the destruction of the lower end of the latter it is made to terminate in a sort of annular inverted grate, having outer and inner water cooled rings E and F. Inlet and outlet water circulating pipes are arranged as at R. The coal container D is fitted with vertical division plates which divide it up into sections, each one of which has its own coal charging inlet as at G. Openings are provided at I from the central air inlet pipe K into these divisions so that a portion of the incoming air passes down through the fuel and carries the distilled hydrocarbons forward to the fire. This method of introducing the fuel and air to the upper fire is adopted in order to obtain as much uniformity as possible in the feed of fresh fuel. The second air inlet H is connected with an external vapouriser in which the steam for working the producer is raised by the heated gases passing off at A. M, N, and P are small inlets with easily removable covers and are used for inspecting the condition of the fire from time to time and also for poking.

The bottom generator is constructed with a revolving bottom of similar form to that described in Chapter 6, p. 90. The fuel rests on a cast iron circular sole plate S, above which is the conical coal hopper T, and there is just sufficient space between the two to allow of the removal of ash and clinker. The circular sole plate S is mounted on balls, and is provided on its underside with teeth which engage with the pinion connected with the outside gear, by which it can be easily turned a few times a day to shake down the ash and clinker from the interior of the producer. The air and steam supply are arranged at L and pass forward to the central trunnion pipe, from which they emerge under the conical cap, which prevents the fuel from filling up this central pipe.

In starting this plant, the lower generator is got going in a similar manner to that employed with the usual simple producer, and coke or non-bituminous coal is used to avoid the nuisance of dense smoke, which would leak out of the various doors and openings. When the lower fire is burning well, the generator is filled up to the coal container in the top portion of the producer and oily waste and shavings are then introduced into the vertical partition in the container, being afterwards ignited to get the upper fire going. After a while both the upper and lower fires will be sufficiently bright to produce a suitable gas at the outlet A, and the plant may then be put to its work.

Figs. 50 and 51 shew the general arrangement of such a plant of 400 B.H.P. in which the producer is worked on the suction principle by means of an exhausting fan discharging through scrubbers into a gas holder. Referring to the plan view in Fig. 51, A is the generator, B the external vapouriser, C the exhausting fan, D a cooler, E and F are scrubbers, and G is the gas holder.

We have described this interesting plant in considerable detail to shew how the principle of fixing or partially fixing the tarry vapours which distil from semi-bituminous fuel by passing them through a hot zone is carried out. It will be seen, however, that the producer is very expensive in construction, and it cannot but require considerable judgment and skill if it is to be successfully operated. The scrubbing apparatus would appear to be considerably more extensive than that which is employed with non-bituminous fuel, and it would certainly seem reasonable to suggest that equally good, and certainly more reliable results, would be obtained by a simple producer with single combustion and the addition of an extra purifier to remove that portion of tarry matter which is now converted into fixed gases in the producer.

General Conclusions. So far as this country is concerned, there is no great difficulty in obtaining a ready supply of suitable gas coke at prices under fifteen shillings per ton, and with a non-bituminous fuel available at such a price it would clearly be unprofitable to attempt to use bituminous fuel for small and

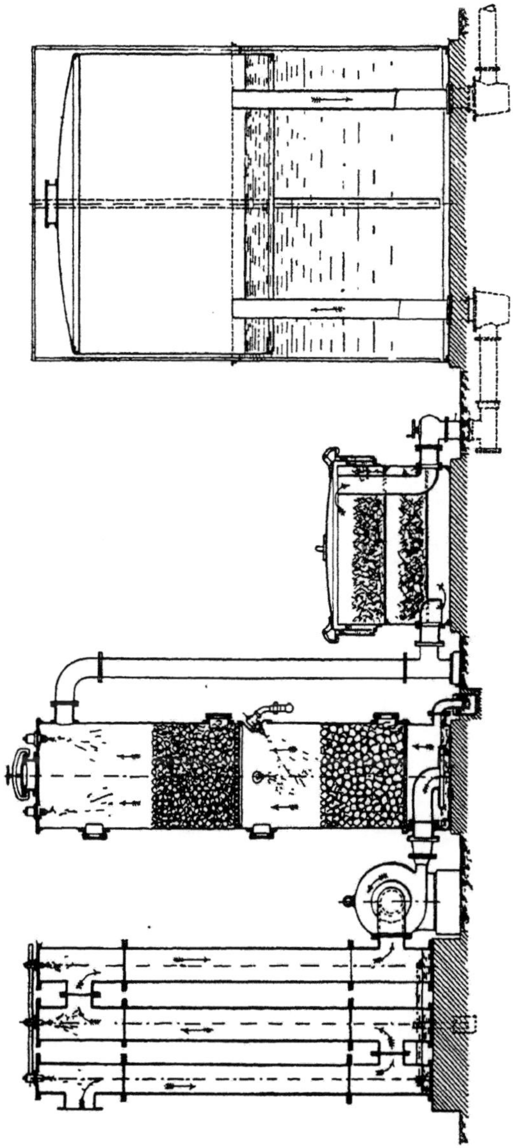

FIG. 50. Section through coolers and scrubbers used with Fichet and Heurtey plant.

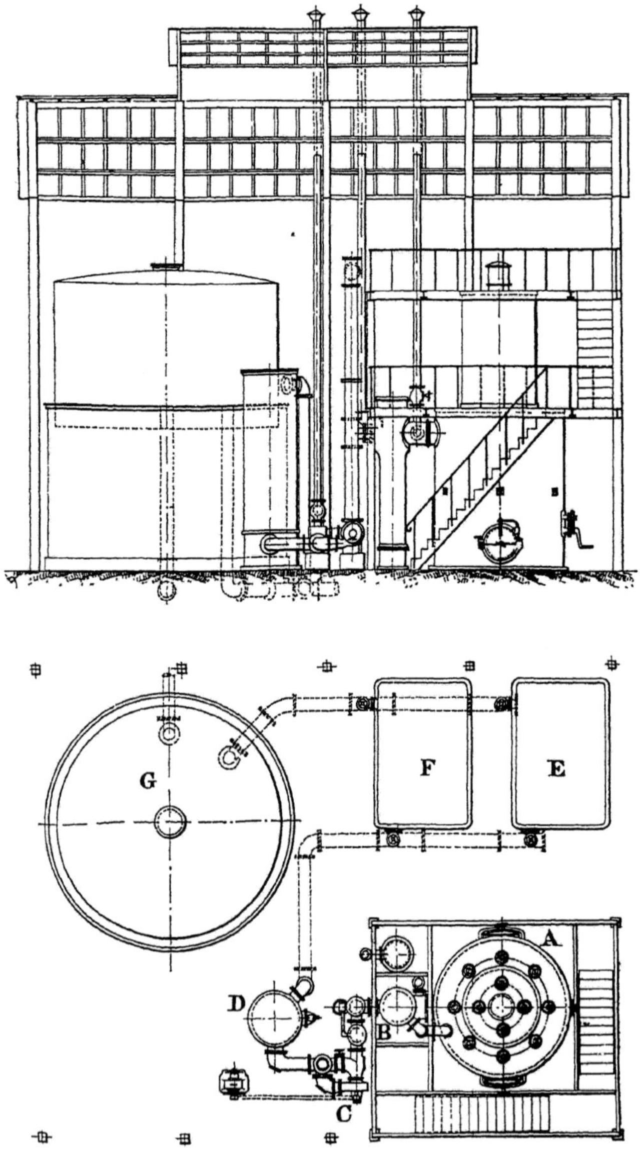

FIG. 51. General arrangement of 400 B.H.P. Fichet and Heurtey plant.

moderate sized plants with the more costly apparatus and increased amount of attention entailed. By far the greatest demand amongst individual users is for power units up to 100 B.H.P. capacity, and the attention necessary to work suction plants of this power is equal to not more than half a man's time over the working week when non-bituminous fuel is used: in plants under 50 H.P. less attention even than this will suffice.

The advantages of a cheaper fuel, therefore, are greatly discounted if its use involves at least double the cost for attendance together with greatly increased first cost and a much more complicated plant. The author has been concerned in numerous installations of gas engines and suction plants of over 500 B.H.P., where with a convenient supply of good gas coke, or Scotch anthracite, the all-round economy of working has been much better than was possible had plants for use with bituminous fuel been adopted. For such reasons it is safe to say that for a power demand up to 300 B.H.P. the ordinary suction plant for use with coke or anthracite can be proved to provide the cheapest source of power in the great majority of cases, and even for installations above 300 B.H.P. most careful comparisons of working costs require to be made in each individual case before it can be laid down with certainty that the advantage will lie with the use of bituminous coal producers.

On the Continent, however, the price of Gas Works coke is high and the supply restricted, which remark also applies to English and Scotch anthracite. Increasing attention therefore is being given to the utilisation of semi-bituminous anthracite there available. We consider that both as a matter of general design, and also for convenience of manipulation, there are many good points in working on the down-draught principle, which are likely to be developed with practical advantage in the future.

With ordinary up-draught producers it is becoming customary in French practice to largely increase the capacity of the coal containers, so that it may hold a ten hours' supply of fuel when running at full load. The object is to raise the temperature of the fresh fuel before it passes to the hot zone, so that the volatile hydrocarbons which chiefly form tar are distilled off, and to some extent gradually forced through the hot fuel and fixed before they pass off with the outgoing gas. There is something to be said for this idea, which, coupled with the larger furnace volume for the duty to be met, will undoubtedly make an ordinary type producer much more suitable for inferior fuels. Larger producers and larger scrubbers, together with some simple but effective tar extractors and sawdust purifiers, will meanwhile go a very long way towards making an efficient apparatus for use with the fuels to which we have referred. Some French makers have also introduced a chemical purifier containing oxide of iron for the purpose of eliminating sulphur compounds from the gas before it passes to the engine, but this additional complication is rather difficult to justify in relation to the actual advantages it confers.

Another important consideration with reference to bituminous coal producers has recently arisen. The bad general effect of the black smoke nuisance in all our large towns is being increasingly realised, and public opinion is quickly tending to the sanction of its absolute prohibition. With this in view, various commercial undertakings are now being formed to treat bituminous coal in bulk so as to extract the volatile hydrocarbons on such a scale that they can be profitably treated, the residual coke constituting a smokeless fuel suitable for domestic and industrial purposes. With a ready supply of cheap coke, the need for bituminous coal gas plants as we now understand the term almost entirely disappears. The treatment of this class of fuel in the manner described would appear to us to embody a sound principle which will sooner or later be made commercially successful.

CHAPTER 8.

WORKING AND MANAGEMENT.

It is of the utmost importance that those concerned with the application of gas power should be perfectly familiar with all essential practical details upon which successful working depends. We find as a matter of common experience that where the introduction of gas power has not proved a success, the cause of complaint is usually found to arise from improper management. In the case of a steam plant, if the attendant omitted to set the feed pump or injector working at the proper time, thus causing the boiler to run short of water and the furnace crowns to come down, this would not be considered to be the fault of the steam plant, but of the attendant. Similarly in the case of a gas plant or gas engine, if through ignorance or inadvertence some equally vital point is overlooked in the general method of working, a careful differentiation must be made in deciding where the responsibility lies before attaching blame to the apparatus rather than to the character of the attention given to it. On the other hand we have invariably found that where power users have realised the necessity of acquiring something like equal knowledge of their gas power plant as they have previously had of their old steam plant, the results they obtain from the former are most gratifying. For such reasons we have given in detail, in the present chapter, proper instructions for the every-day working and management of suction plants, and though for clearness only one particular plant is described, it will be understood that the general methods laid down apply to all types of plants.

Special points to be observed in new plant. When putting a new generator into use for the first time a small fire only should be lit in the furnace, so that the new brickwork may have a proper opportunity of becoming thoroughly dried before being subjected to the high temperature met with under working conditions. In the absence of this precaution the brick work will probably be cracked during the first run and the serious trouble mentioned in Chapter 4, p. 54, will result.

Another point to be carefully remembered in new plants is that the pipes and scrubbers are full of *air* to start with. This air must be expelled before any open flames are applied to any of the test cocks fitted to or connected with the scrubbers on gas main. As soon as a good burnable gas is being made at the producer, the blow-off pipe at the expansion box may be opened and the chimney cock shut, and thereafter gas should be steadily blown through the scrubber and gas main *for at least five minutes* before attempting to start the engine or to apply any flame to the test cocks on the gas main near the engine. The reason of this precaution is that a mixture of gas and air in the scrubber will explode with possible serious results if a light is applied to any of the connections named. All traces of mixed gas and air must therefore be expelled, and when only gas remains no danger may be feared.

INSTRUCTIONS FOR STARTING AND WORKING. (See Fig. 52.)

Fuel to be used. For small plants it is best to fire the gas generator with anthracite peas or beans in pieces about ⅜″ to 1″ cube. For larger plants anthracite, about 1¼″ to 2″ cube. If suitable anthracite cannot be obtained, gas coke may be used if clean, of good quality, and in pieces about ½″ to 1″ cube. The names of suitable firms from whom these fuels can be obtained will be given by any of the gas plant makers.

Preparing the plant for work. Gas generator. When starting up from cold it should be firstly seen that the blow-off cocks on the chimney pipes and from the expansion box can be freely turned, also the test cocks on the chimney pipe and near the expansion box. It is impossible to prevent these cocks from sticking at times, and as they are only used on starting up, it should be seen that they are quite free before commencing to blow up the fire.

To start the fire in the generator, place on the grate near the fire door a handful of cotton waste soaked with paraffin, then drop in wood in short lengths through the hopper, also a bucketful of coke ¾″ cube. Lower the hopper valve on its seating, and it should be seen that this latter is clean so that there is no doubt of the valve making an air-tight joint. Open the cock (14) on the blow-off pipe, close the disc valve (6) on the air inlet pipe, and turn on the water to the generator and coke scrubber by opening the water valves (22) and (23) respectively. It is essential for producing good gas at starting that the bottom of the ashpit should be covered with a layer of water, and it is best to give the ashpit bottom time to fill up by the water overflowing from the vapouriser through the down pipe (10) before attempting to light the fire. When water is observed in the ashpit, the oily waste should be set on fire and the fan turned. After turning for about a couple of minutes, it is best to make

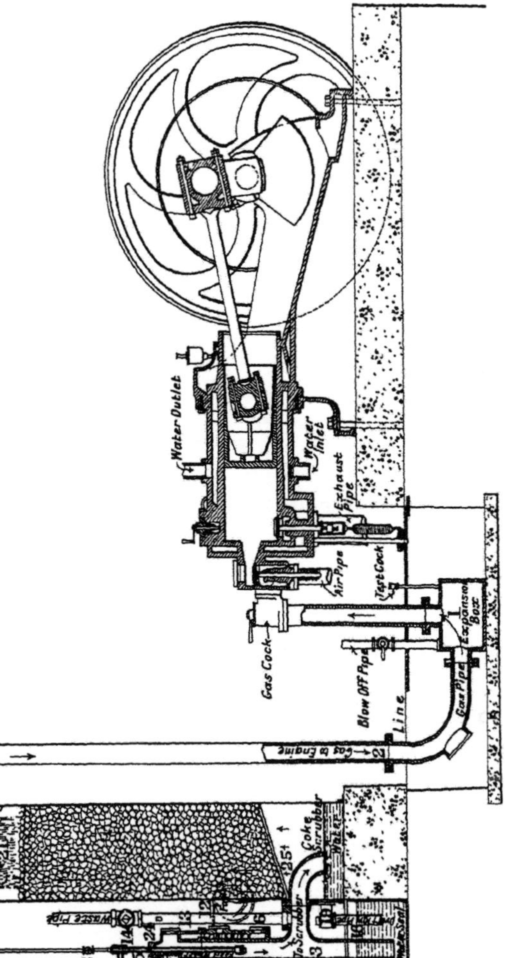

FIG. 52. Section through a suction plant and suction gas engine.

certain that the wood has caught fire properly, the fan being stopped and the fire door opened for a few seconds to ascertain this.

As soon as the fire has started properly the generator may be fully charged right up to the hopper valve, and the fan should then be turned steadily at a speed of about 40 turns per minute for about 10 to 20 minutes, until the burnable gas is obtained at the test cock (24) on the chimney pipe.

When the gas burns at this test cock, open the blow-off cock on expansion box near engine, and close the outlet cock on chimney pipe. MAKE SURE THAT THE ENGINE IS IN THE CORRECT POSITION FOR STARTING WITH ALL THE VALVES CLOSED; OTHERWISE GAS MAY PASS THROUGH THE GAS AND AIR VALVES, AND ESCAPE INTO THE ENGINE ROOM. After a few minutes, test the gas near the engine, and when it burns well start the engine. Open the test cock only when testing the gas and shut it immediately afterwards.

AS SOON AS THE ENGINE IS RUNNING AT FULL SPEED, close the blow-off cock from expansion box, open the disc valve (6), which admits air to the generator, cease working fan, and close the air inlet cover hinged on to the air inlet to the fan.

The generator should be filled with fuel so as to keep the level after each charging operation to the height shewn on the sectional drawing above, and while the engine is at work the fuel must never be allowed to get down below the level of the fuel container. In charging the generator with fuel, the cover valve (17) must always be in position before attempting to open the lift valve (16), and the latter should always be pressed on its seating by working the handle to and fro a few times after a fresh charge of fuel has been put in.

The essential points, therefore, in working the generator are:

(1) TO USE THE PROPER SORT OF FUEL IN ACCORDANCE WITH THE INSTRUCTIONS ALREADY GIVEN.

(2) TO KEEP THE GENERATOR PROPERLY CHARGED WITH FUEL BY FILLING IT UP TO THE PROPER LEVEL AT REGULAR INTERVALS, ACCORDING TO THE LOAD.

(3) TO SEE THAT THE PROPER SUPPLY OF WATER FOR MAKING THE NECESSARY STEAM IN THE VAPOURISER

IS RUNNING IN AT THE PROPER REGULATION, WHICH IS AT THE RATE OF $1\frac{1}{2}$ GALLONS PER HOUR FOR EACH 20 B.H.P. DEVELOPED BY THE ENGINE; OTHERWISE THE GENERATOR SHOULD NOT BE INTERFERED WITH WHILE AT WORK.

Shutting down Gas Plant between Working Hours.

To stop the engine, shut gas cock on engine, and immediately afterwards open the cock on chimney pipe of generator.

During stoppage at meal time or at night, leave the cock on chimney pipe open. On re-starting, fill up with coal and commence blowing with fan as before.

In the morning loosen the clinker by working a bar across the top of grate and down through the hopper. Rake out clinker, replace the doors, fill up the fuel container, and blow up the fire till the gas is good enough to start the engine. The fire in large generators should be drawn once a week, and the generator thoroughly cleaned out.

With small generators, up to 20 B.H.P., it is best to draw out all the fuel from two or three times per week. The anthracite taken out of the generator should be screened through a riddle having holes about $\frac{1}{4}''$ square, the pieces of clinker, etc., being thrown out. It should then be mixed with new anthracite and then used again in the generator during the day.

The drain pipe from the bottom of generator should be examined every morning, after clinkering or drawing out the fire, to see that it is quite free from obstruction.

Shutting down Gas Plant for cleaning out Generator.

All generators should be carefully cleaned out at the end of each week's run, as there is always an accumulation of clinker inside of the furnace as a result of the week's working, and which can only be removed by taking all the coal out of the generator and detaching the clinker from the sides of the brickwork.

It is important that the fire should not be drawn immediately after the plant has stopped working and while it is consequently very hot. If this be done there is a great risk of cracking the brickwork and the shell of the generator through too rapid contraction.

The air inlet to the fan and also the disc valve (6) should therefore be closed some hours before any attempt is made to withdraw the fire. The chimney cock (14) should be left open to carry off any accumulated gas, but the result of closing all the air inlets to the furnace is that the fire gradually dies out, and it may be drawn at any time thereafter without risk to the plant.

Coke Scrubber.

The coke scrubber should be filled with clean coke in pieces about 2″ cube for the first 12″ from bottom, and thereafter with pieces ¾″ cube. The whole of the coke in the small scrubbers with one grid should be taken out and renewed every three months; where there are two grids the coke in the lower one should be renewed once a month. The coke on the upper grid should be changed once a year. When there are two scrubbers, the coke in the second one should be changed once a year. So much of the coke taken out as is clean and of good quality can be mixed with anthracite, and if in pieces ½″ to 1″ cube used in the generator. **See special cautions below as to opening and cleaning scrubber.**

During the working of the plant, sufficient cold water must be passed through the scrubber to cool the gas, which should be cold when leaving the scrubber. From the scrubber the water will pass into the overflow tank.

The drain pipe from bottom of scrubber must be set so as to keep the proper level of water in bottom of scrubber. Referring to Fig. 52 above, it will be seen that the gas outlet pipe (3), between the generator and the scrubber, connects to an internal bent pipe (25) which dips ¼″ into the water at the bottom of the scrubber. It is essential that it should not dip in more than this ¼″ and consequently the down pipe of the overflow (19) must be kept free from obstruction, and its height adjusted to keep the correct level of water in relation to the bent pipe (25) at the bottom of the scrubber.

DETAILS REQUIRING REGULAR ATTENTION.

Gas generator. If it should be found when cleaning the inside of the generator that there are any considerable cracks in the brickwork these should be carefully filled up with good fire-clay or ganister.

If the firebars appear to be badly burned, this is due to working the plant with insufficient steam and consequently the temperature of the fire has been too high.

If the top of the plant and the outlet pipe become very hot during working, it is probable that an air leak into the inside of the generator is taking place either through the top joints or through some of the brickwork. This should be remedied immediately.

Soft water should preferably be used in the vapouriser of the generator, so that difficulties on account of scale forming can be avoided.

The chimney cock sometimes becomes made up with tar and soot and as a consequence there is no vent when blowing up with the fan. This makes it impossible to get a good gas for starting the engine and consequently this cock and the chimney pipe should be regularly cleaned.

SCRUBBERS.

See special instructions above as to the charging of the scrubber in the first instance, and the renewing of the coke at intervals.

The base of the scrubber should be cleaned out every fortnight. It is especially important to keep the end of the dip pipe and the top of the overflow pipe clean, and to remove any deposit on the inside of the bend of the dip pipe, also see that all the dirt is removed out of the overflow tank (18) every fortnight.

For working with gas coke, a tar extractor and sawdust scrubber are sometimes added to the ordinary coke scrubber. The sawdust should be renewed every five or six weeks when the plant is running on full load, and the dash plates in the tar extractor should be cleaned every week, the tar being scraped or burnt off as required.

Fan.

Oil must be kept in the bottom of casing enclosing the gear wheels. An oil filler is provided on the casing for filling up, and the oil should be kept right up to the level of this filler, and a supply of oil should be put in once a week for this purpose.

Once a day a few drops of oil should be put on the end of the spindle carried by the bridge piece at the air inlet.

The starting and stopping of the fan should be done slowly, and the handle should not at any time be turned more than 50 revolutions per minute.

Cleaning of Apparatus.

Examine all gas mains, connections, cocks and valves from time to time, especially those near the generator, and see that they are clear of the condensed moisture and deposit. This also applies to the expansion box near engine, which should be drained by removing plug once a week.

Special Cautions.

The gas made in suction producers is poisonous. It has little or no smell, and it should not on any account be allowed to escape into the engine-room or any other place where it might be inhaled. An escape can only take place in the usual way by the cocks or valves being left open while blowing with the fan, and before starting up the plant all cocks and valves on engine and plant should therefore be carefully examined to see that they are shut or open only in accordance with the instructions given above.

Do not take out any cleaning plugs or open any bolted-on doors while blowing with the fan, or while the plant is at work.

The blow-off pipes from the generator and expansion box should be led into the open air clear of any windows or other openings through which the gas might pass back into the building.

When the producer is being used for the first time after erection the blow-off cock from the expansion box should be left open for at least five minutes during blowing, so as to expel the air in the pipes and scrubbers, otherwise an explosive mixture might be formed therein which will be fired in trying the gas at the test cock.

After a scrubber has been used no person should enter it until the doors and top have been open for several hours to allow the lurking gas to escape, and no naked light should be brought near a recently opened scrubber.

Special notice must be taken of the instructions for charging the generator, as it is essential that the feed valve and the top cover valve should never be open together, and no attempt must be made to look inside the generator from the top either while the plant is working or immediately after the engine has stopped.

Working of Gas Engines with Suction Plants.

(1) The gas should be quite cool when it reaches the engine, and if it is not so it is fairly certain that either the supply of cooling water to the scrubber has failed or is too much diminished, or that the sprayer inside of the scrubber has become choked.

(2) The engine valve spindles should be thoroughly cleaned and kept quite free, and the valves ground to their respective seatings at least every three or four weeks or more frequently if a dirty coal is used.

(3) The ignition plugs should be thoroughly cleaned and overhauled every week if circumstances permit.

(4) Back firing is usually caused by a combination of the following circumstances, or by either individually:

(*a*) Dirt in the engine cylinder.

(*b*) Irregular or weak ignition spark.

(*c*) Too much air, or conversely too little gas in the supply of mixture to the engine. This is overcome by correctly adjusting the gas and air inlet cocks or valves.

(*d*) Weak gas caused by too much or too little steam or by a stoppage in some of the passages of the gas plant. If any of the important joints of the gas plant give out, bad gas will also result. (See Chapter 4, p. 52.)

(*e*) Some passage or flaw in the combustion chamber casting of the engine, which holds the flame over from the last power impulse and ignites the incoming charge.

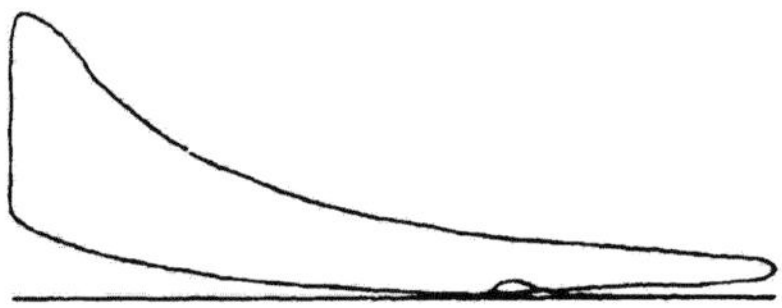

FIG. 53. Indicator diagram shewing a 'back-fire' during suction stroke.

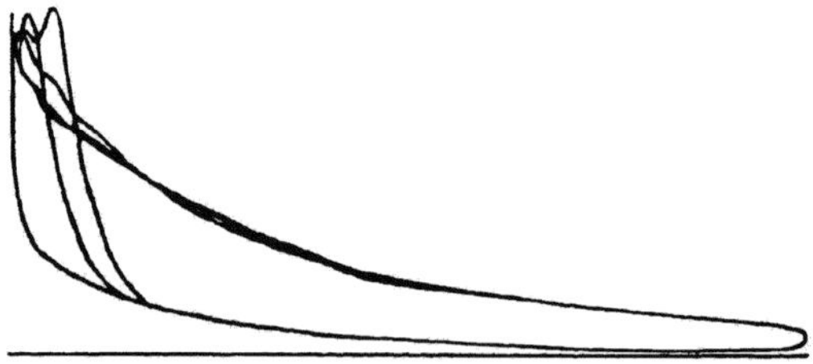

FIG. 54. Indicator diagram shewing 'pre-ignitions' during compression stroke.

(5) Pre-ignitions are produced by dirt in the cylinder, and also by cause (*e*) in No. 4 above. Figs. 53 and 54 represent indicator cards shewing a back fire and pre-ignition respectively.

(6) Missed ignitions usually result either from wrong proportion of gas and air in the mixture supplied to the engine, from weak gas or from dirty contacts on the ignition plug.

The remedy for any one of these is obvious.

(7) Failure to get the gas through from the plant to the engine whilst working fan for starting purposes is due to a stoppage in the gas passages at some point—usually the water seal (Fig. 37) has become choked. The exact point of the

stoppage can in any case be located by an intelligent use of a U shape water gauge as in Fig. 18, Chapter 3. The engine should be indicated from time to time and both power and

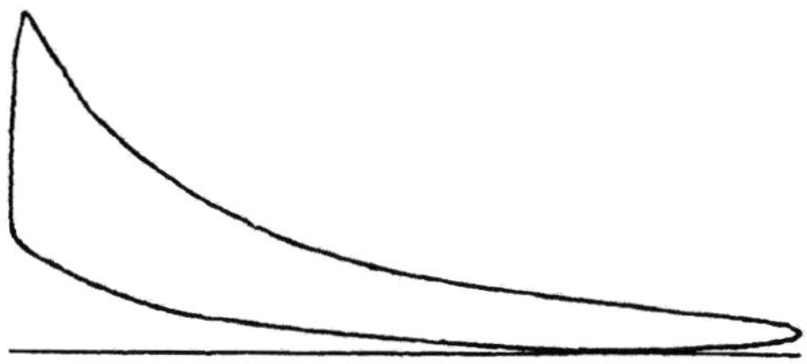

FIG. 55. Normal indicator power diagram.

FIG. 56. Defective indicator power diagram due to bad gas or wrongly adjusted mixture of gas and air.

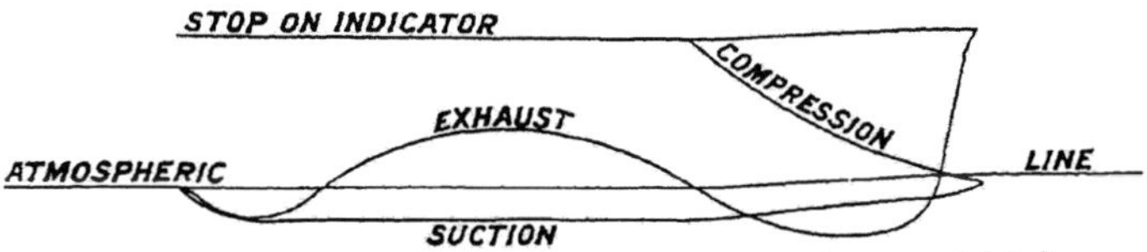

FIG. 57. Normal light spring indicator card shewing unrestricted exhaust and inlet passages.

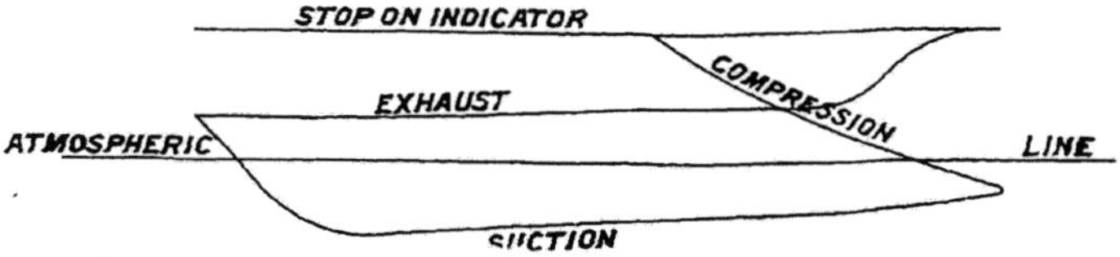

FIG. 58. Light spring indicator card shewing effect of restricted exhaust and inlet passages.

light spring cards should be taken. Figs. 55, 56, 57 and 58 above shew normally good cards, and also cards taken where defects exist.

Poisoning by Producer Gas.

With suction plants the chances of poisoning due to an escape of gas are very small indeed, as leakages while the plant is at work are quite out of the question. If by inadvertence, however, the gas should be inhaled whilst cleaning scrubbers or whilst blowing up with the fan, it is advisable to understand the steps which should be taken:

Effect the gas produces. The first *symptoms* produced by breathing the gas are giddiness, weakness in the legs, and palpitation of the heart.

If a man feels these symptoms he should *at once* move into the fresh air, when in slight cases they will quickly disappear. Exposure to cold should be avoided, as it aggravates the symptoms.

A man should not walk home soon after recovery, as muscular exertion when affected by the gas is to be avoided.

If a man should be found insensible or seriously ill from the gas, he should at once be removed into the fresh air, kept as warm as possible, and oxygen administered, a medical man being sent for at the same time.

To use the oxygen cylinder. Open the valve gradually by tapping the lever key (fully extended) with the wrist until the oxygen flows in a gentle stream from the mouthpiece in the patient's mouth and allow the oxygen to be breathed until relief is obtained. The lips should not be closed round the mouthpiece, as it is important to allow free egress for surplus oxygen. The nostrils should be closed during inspiration or inflation of the lungs, and opened during expiration or deflation of the lungs, so that the oxygen may be inhaled as pure as possible through the mouth. *If the teeth are set*, close the lips and one nostril. Let the conical end of the mouthpiece lightly enter the other nostril during inspiration and remove it for expiration.

Artificial respiration. Artificial respiration is sometimes necessary in addition to the oxygen inhalation, if the oxygen does not appear to act quickly.

Place the patient on his back, slightly raising the shoulders with a folded coat; remove everything tight about the chest and neck; draw the tongue forward and maintain it in that position. Grasp the arms just above the elbows, and draw them steadily above the head, keeping them on the stretch for two seconds and then folding them and pressing them against the chest for the same length of time. Repeat these movements about 15 times a minute for at least half an hour or until the natural breathing has been initiated, when the oxygen inhalation will alone suffice.

(After recovery, oxygen inhalation at intervals should be continued as desired.)

No man should work alone on any work which would be likely to involve exposure to the gas. Should circumstances cause a man to enter a scrubber or generator where gas may be lurking he should have a rope tied round his waist, held at the other end by his mate standing outside.

CHAPTER 9.

APPLICATIONS TO PRACTICE.

Combinations of suction gas plants and gas engines are being applied in large numbers for all purposes, and their reliability and economy have been practically demonstrated in thousands of cases where they have been at work for several years. As a cheap form of motive power they are unrivalled, and are used for driving mills and factories, shipyards, bridge works, tin and copper mines, coal mines, weaving sheds, spinning sheds, saw mills, flour mills, laundries, cold stores, cement works, sewage and pumping schemes, farm and estate work, etc. etc.

In electrical work they are extensively used for theatre lighting, lighting of small towns, mansions and country houses, driving of large electrical generators for power purposes generally, electro-chemical processes, etc.

For producing cheap current, it is usually considered that no Public Supply can compete with a private Gas Power installation properly arranged and worked.

Special points in applying suction gas power. It was at first objected that in those instances where steam pressure was in any case required for driving power hammers or for heating warming pans, the economy of the gas would be seriously reduced as a separately fired boiler for the purposes named would in any case be required. So far as hammers are concerned, the difficulty is easily and efficiently solved by using compressed air instead of steam, the compressor either being combined with the hammer and driven by belt from the line shaft, or else a separate compressor is put down for the purpose. Compressed air is found to be a much better medium of power than steam for this purpose, and apart from gas driving altogether, air hammers are being increasingly favoured.

Where small quantities of steam are required for heating purposes, it is frequently found that a sufficient supply can be obtained by passing the exhaust gases from the engine through a special steam raising boiler. It must be borne in mind however that these boilers are only efficient as steam raisers when the engine working in conjunction with them is running at from three-quarters to full load. At light loads, where the explosions are much reduced, either in number or intensity, a considerable volume of hot water can be obtained from the waste heat, but not much steam. Otherwise when the load is sufficient as stated above 2½ to 3 lbs. of steam at 60 lbs. pressure can be obtained per hour for each B.H.P. developed by the engine. The boiler must be placed as close as possible to the engine. The Wilson Patent Boiler made by Barclay & Co. of Kilmarnock has proved to be efficient for this class of work, and a section of same is given in Fig. 59 and a diagram shewing how it is usually connected with the engine in Fig. 60. A bye-pass for the exhaust gases is usually provided so that they may be passed direct to the atmosphere without passing through the boiler if desired. In flour mills, cement works, saw mills and chemical works, the utilisation of the heat of the exhaust gases, which would otherwise be lost, will be found of special advantage.

Limit of economy as compared with town gas for engine work. In certain of the large towns in England gas may be obtained for power purposes at very low rates. In

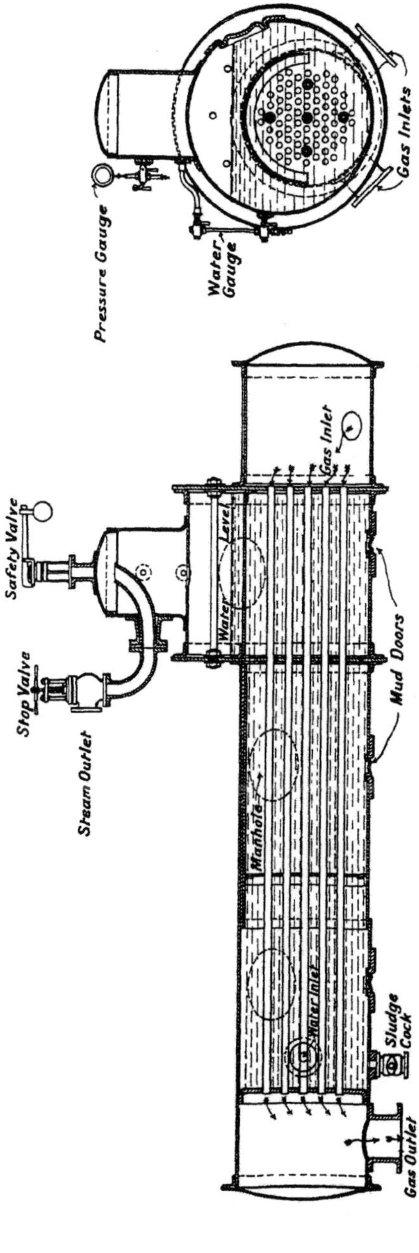

FIG. 59. The 'Wilson' exhaust heat boiler.

Sheffield for instance the price is as low as 1*s*. per thousand cubic feet. It may be taken in such cases that up to 100 H.P. units, there is little all-round gain in applying producer gas unless the engine is required to run continuously at full power. For instance with a 100 H.P. engine, the average load is not

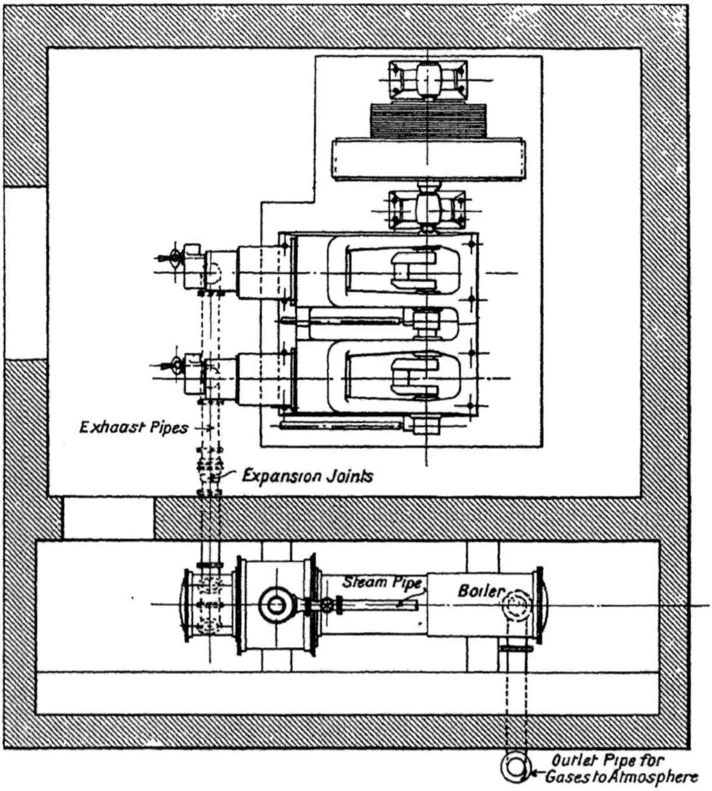

FIG. 60. Diagram shewing the method of connecting an exhaust heat boiler to the engine.

more than 50 H.P. over the working week, the sum of the gain which will accrue by using power gas is much less than if the engine is running continually at full power. The proper claims of suction gas power are so numerous and well founded that cases of doubtful economy following on its application, through gas being obtainable at low rates from the Public Supply, may well be left alone by those interested in the sale of combined installations of engines and producers. All such doubtful cases should be carefully looked into on their merits.

Difficulties arising from widely fluctuating loads. Due to the lag of the fire in the producer (see Chapter 10, p. 146), a difficulty has sometimes arisen in working suction producers satisfactorily where the load rapidly fluctuates over a considerable range. The difficulty is best got over by fitting an engine and producer amply large for their work, and in extreme cases engines of the so-called electric type with heavy flywheels may with advantage be adopted. The momentum of the heavy flywheel acts effectively as an equaliser and keeps up the speed of the engine even if the quality of the gas falls away through a heavy load being suddenly applied. It may be stated with confidence that makers of experience can now meet successfully all cases which arise in practice where fluctuating loads have to be dealt with.

Working several engines from one suction gas plant. There is no difficulty in arranging this, but a separate gas main should be brought direct from the scrubber to each engine. If a common main be used with branches therefrom to the several engines it is found that the suction of one engine may interfere with the gas supply to another, especially if the sizes of the units vary considerably. A little difficulty may be experienced in starting those engines furthest away from the plant whilst the others are running, but this is easily got over by bringing the hand blower into use and forcing the gas through the main for the time being.

Suction plants for combined power and heating. Cases frequently arise where in addition to the gas required for driving the engine, it is desired to use the Producer Gas for lacquering stoves, brazing, brass melting, annealing, laundry work, etc. The leading makers have adopted for this purpose an arrangement consisting of an exhauster driven by the engine working from the same producer, which draws off the required amount of gas from a separate outlet in the scrubber. The gas is then forced into the small gas holder weighted so as to give the required gas pressure. As the bell of the gas holder raises, and when it has reached a predetermined height it operates by suitable levers a valve placed in the pipes between the exhauster and the holder, and closes it until only sufficient gas is passed into the holder to make up for that taken away for the purposes required. If no gas is wanted the valve is entirely closed and the exhauster simply churns the gas contained within itself and does not draw on the scrubber.

An arrangement of this sort requires to be applied with some discretion, and the relative demand on the gas plant caused by the engine as against the amount of gas required for heating must be carefully taken into account. If the engine demand is the larger proportion, and its load fluctuates, then the quality of the gas will also fluctuate considerably, and under these circumstances we would not advise the heating and power gas to be taken from the same producer, for should the flames of the gas jets of the heater become extinguished when such a fluctuation takes place and the fact be unnoticed until a considerable quantity of unburnt gas had escaped, gas poisoning or some other undesirable circumstance may occur. Speaking generally we would for heating purposes prefer to apply a plant of the pressure type or a self contained suction plant and exhauster set apart for this work. It is not wise to run any risks of gas escapes in the work rooms where a considerable number of operatives may be present.

Electric driving. Where a number of gas engines coupled to suction producers are assembled in a power house for driving dynamos, it is usually desirable to have a separate producer for each engine, so that each unit consisting of producer engine and dynamo is quite self contained. The various stages of the load curve can then be met under the best conditions. Where the engines and dynamos are of similar size and are working on independent circuits with varying load demands, the gas producers will give more uniform results if worked in parallel.

General. For factory and mill driving, and power schemes generally, the chief consideration in applying gas power is to choose suitable units for the work to be done, and they should be of ample size to comfortably meet the load demand. The difference in the cycle of the gas engine as compared with the steam engine, accounts for the fact that the former cannot be forced, but it must be borne in mind that no engine should be forced and it is a *sine qua non* with gas engines that they should be large enough for their work. This means that when a change from steam or electric driving to the new gas power is contemplated careful computations of the load requirements under all conditions should be made.

Portable gas producers and engines. For farm, estate, and general contractors work, the stationary gas engine and suction producer have proved so great a success that several makers have set to work to provide a portable gas power set. Fig. 61 shews a 13 B.H.P. Tangye combined producer and gas engine which has been arranged specially on the truck so as to keep the centre of gravity as low as possible: the whole apparatus is bolted to a steel girder frame mounted on four large travelling wheels. The gas producer is of the standard type with the exception of an additional air cooler, consisting of a number of atmospheric cooling tubes through which the gas from the generator passes to be partially cooled before entering the coke scrubber, and which consequently has the effect of diminishing the amount of cooling water required to be run through the latter. The generator is arranged at the end of the truck thus enabling it to be readily charged with fuel and generally attended to. A tank of large capacity for holding the cooling water is carried on the girder frame, and the water is forced through the jacket of the cylinder by means of a circulating pump, thence on to a canopy fixed at a convenient height above the engine, which serves as a surface cooler, and afterwards returns to the storage tank. A small quantity of water is allowed to flow into the coke scrubber so as to keep the coke moist for more thoroughly cleaning the gas before it serves the engine.

Applications for marine propulsion. Several types of engines and producers are being tried at the present time on small craft, and a good deal of attention is being given to the subject. Messrs Thornycroft have fitted a gas engine and suction plant to a canal barge and Messrs Campbells of Halifax

Fig. 61. Portable suction engine and gas plant (Tangye).

to a Dutch tug-boat of 100 B.H.P. This aspect of gas power is, however, too much in the experimental stage to allow of serious treatment within the scope of this work.

The general arrangement of gas engines and suction gas producers. We have already mentioned, at the beginning of this chapter, some of the many uses to which suction gas power is being successfully applied at the present time, and we now draw attention to a few typical general arrangements of the complete power installation in such cases.

Fig. 62 shews, what might be considered, a standard arrangement of a 300 B.H.P. power house for driving a weaving or spinning shed. The producer is usually placed in a well-ventilated shed adjoining the engine house so that the attendant can easily look after both producer and engine at the same time. The power is taken off the end of the engine crank shaft from a rope pulley connected to the latter through a friction clutch which facilitates starting, and which affords a ready means of stopping the mill shafting quickly in case of any accident to the machinery which the engine is driving. Compressed air is usually applied for giving the preliminary momentum to the engine so as to enable it to draw in its own charge of gas and air when starting up: a small independently driven air compressor is all that is necessary for this purpose. The method of connecting up gas, air and exhaust piping will be understood from the figure, and the cooling water for the engine jackets is usually obtained from a large circulating tank fitted over the engine house.

Fig. 63 shews a pumping plant for a water works where it is required to move a large body of water from one reservoir to another at times when the balance of supply in the system of reservoirs tends to become disturbed through drought or from any other cause. The pump, which is of the three-throw ram type, is directly connected to the engine through a friction clutch, and pumping may be commenced within half-an-hour of lighting the fire in the generator, starting up from cold. This factor is of first importance in such installation which may not be required for use over many weeks when a demand will come quite suddenly. There are consequently no stand-by

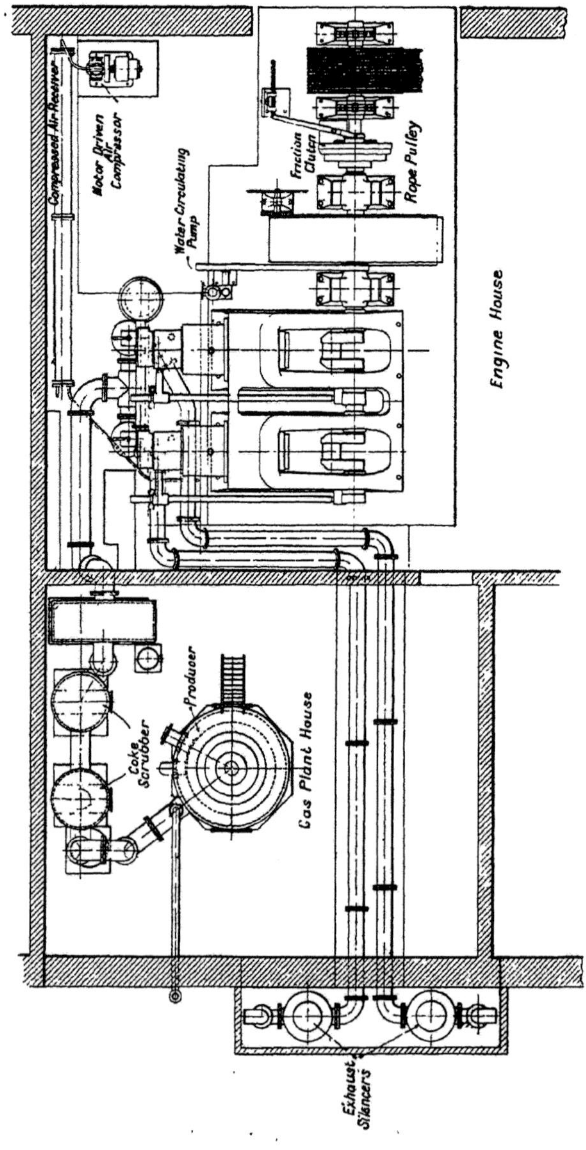

FIG. 62. Arrangement of power house for 300 B.H.P. mill engine and suction gas producer.

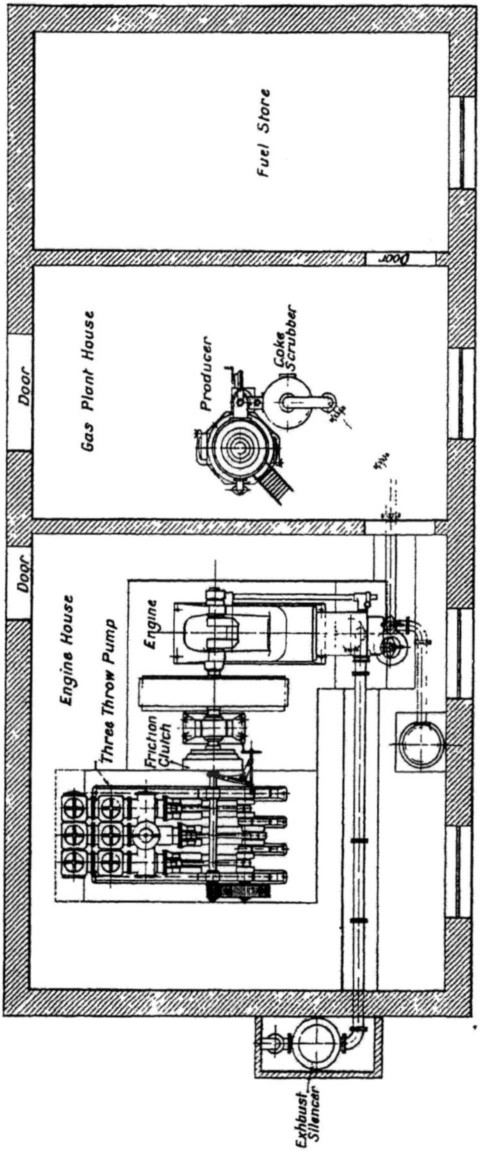

FIG. 63. Arrangement of surface pumping plant of 100 B.H.P. for water works.

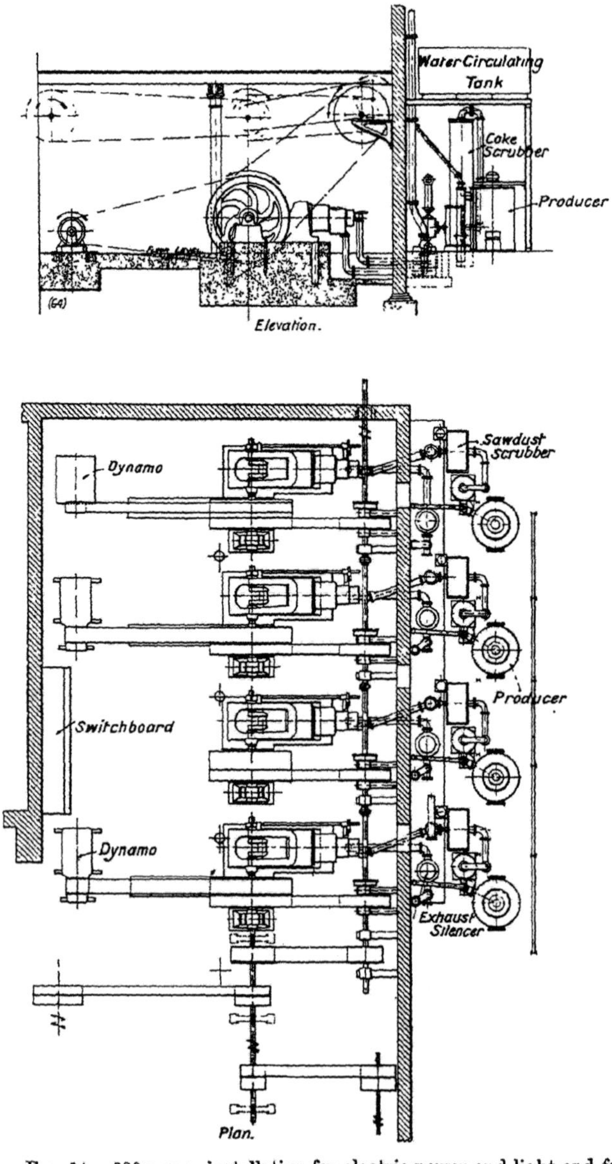

FIG. 64. 600 B.H.P. installation for electric power and light and for line shaft driving.

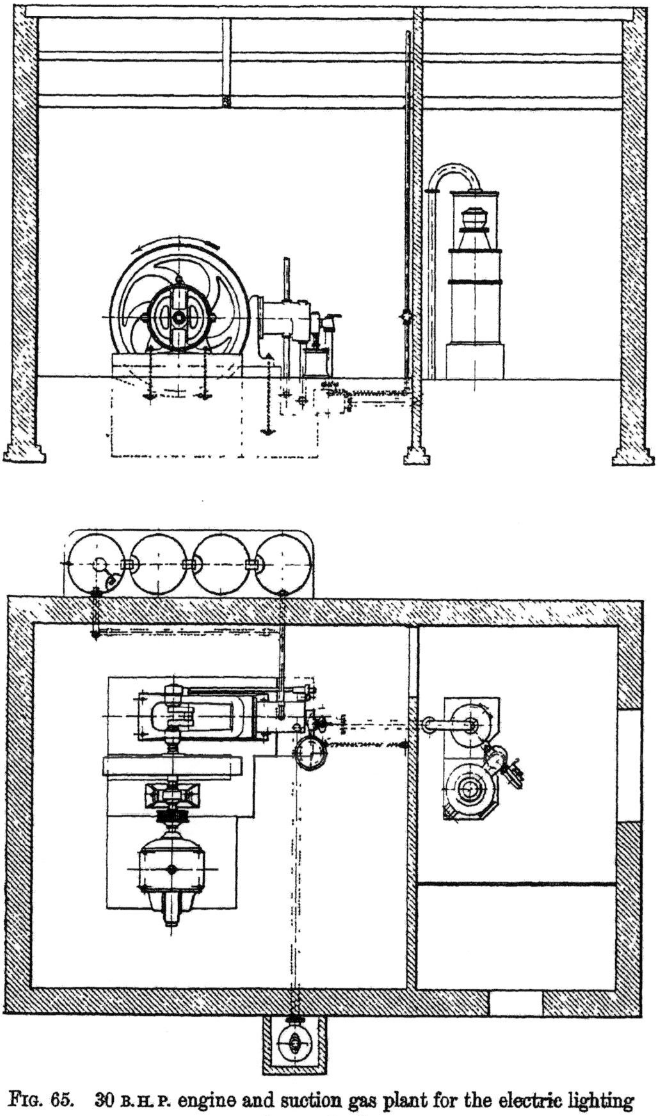

FIG. 65. 30 B.H.P. engine and suction gas plant for the electric lighting of a country house.

losses under these circumstances with a plant of this description, which at the same time is always ready for immediate use. Their great economy, too, is an important matter since water

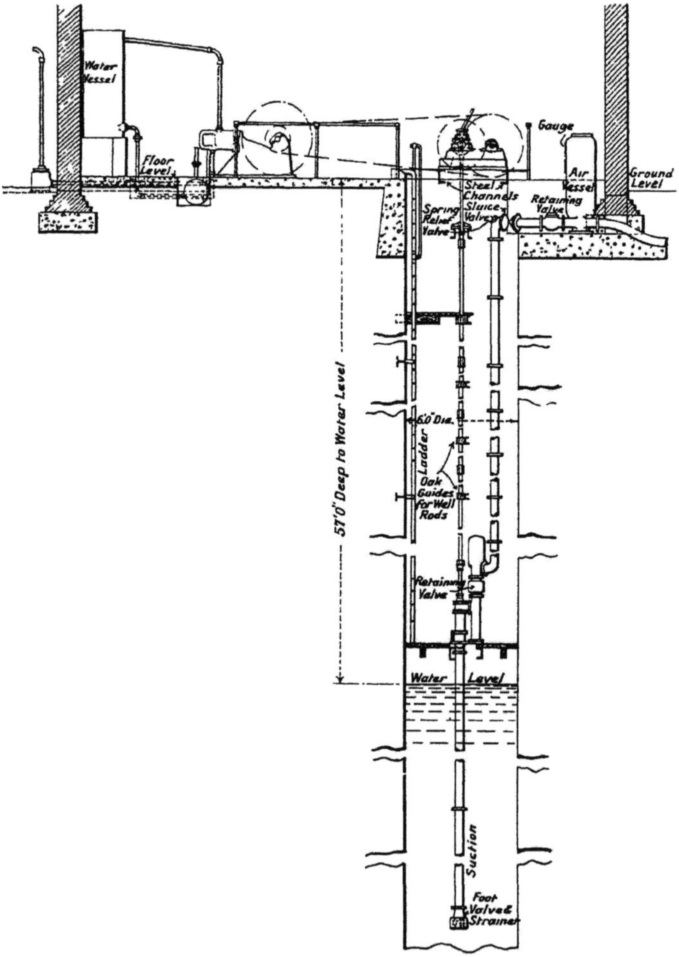

FIG. 66. Sectional elevation of deep well pumping plant operated by suction gas engines (Messrs Tangye, Birmingham).

works are usually situated in out-of-the-way places to which all fuel and stores have to be carted several miles from the nearest railway station.

In Fig. 64, a 600 B.H.P. installation for supplying light and power to a large engineering works is shewn. For elasticity under varying load conditions, four 150 B.H.P. units were installed, each being connected through suitable friction clutches

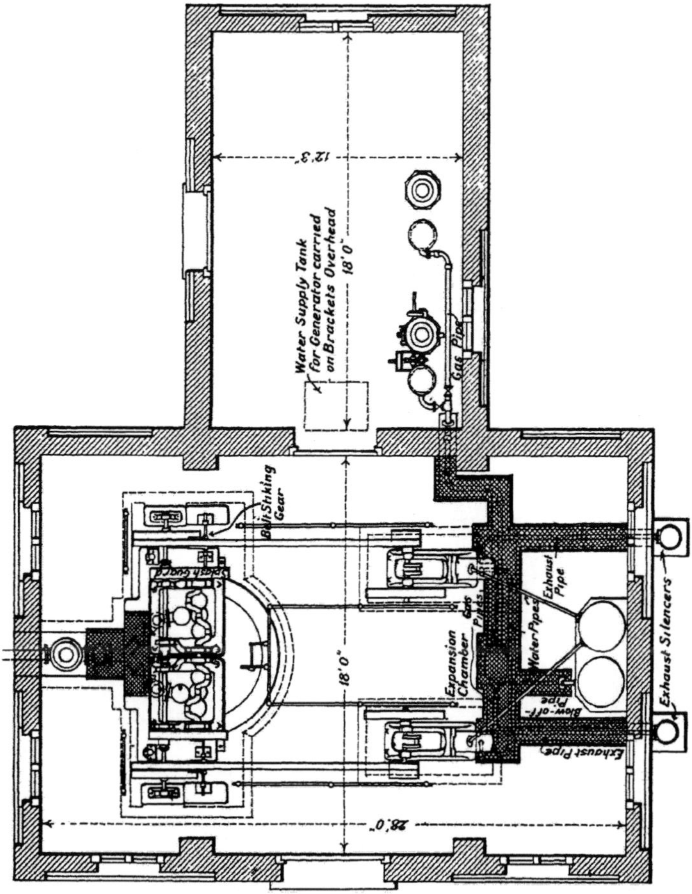

FIG. 67. Plan view of deep well pumping plant.

to the main line shaft as well as being available for dynamo driving. Under working conditions the consumption of Scotch anthracite is $\frac{3}{4}$ pounds per H.P. hour, which of course is remarkably economical, and in other essential ways the installation has proved a complete success.

Small suction gas plants and engines have been very largely adopted for country house electrical installation, for which purpose they have quite superseded oil engines formerly favoured for these isolated small plants. The advantages of a gas plant

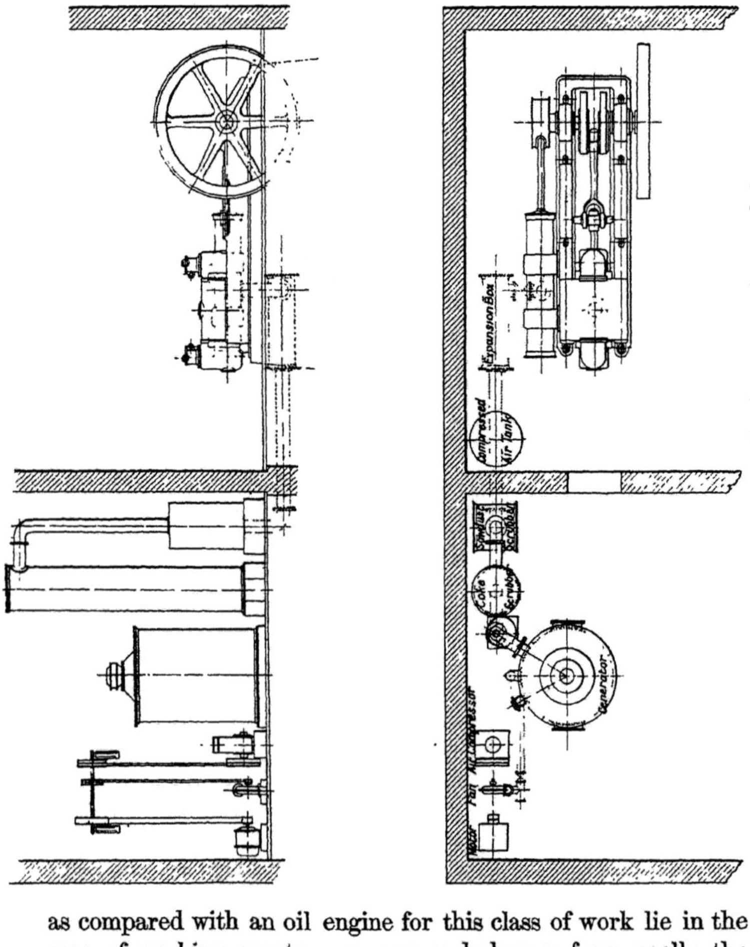

FIG. 68. 250 B.H.P. two-cycle Körting type engine coupled to

as compared with an oil engine for this class of work lie in the ease of working, greater economy and absence from smell; the latter point is particularly important.

Fig. 65 shews a good arrangement of engine and producer house for a country residence. The engine is preferably direct connected to the dynamo, and the accumulator room may be placed over the engine house so that the acid fumes can easily be dissipated. The ordinary cooling tanks for the engine are placed outside behind the engine house.

Figs. 66 and 67 shew in elevation and plan, an interesting deep well pumping installation by Messrs Tangye of Birmingham. The arrangement of the engines, producers, pumps and auxiliary appliances will be well understood from the diagrams. The working economy of suction gas power for isolated installations such as that illustrated cannot be equalled or even approached by any other form of motive power.

Fig. 68 shews a two-cycle Körting type engine by Messrs Mather and Platt of Manchester, coupled to a Dowson suction gas plant. The chief difference between this and the other installations described is that the gas mains are connected to the charging pump at the side of the engine instead of to the working cylinder. We understand that the combined plant here illustrated is of 250 B.H.P. and has been running successfully for some time.

CHAPTER 10.

WORKING RESULTS.

Composition of Suction Gas. We have seen in Chapter 2 how the general composition of producer gas made from carbon is theoretically arrived at. The actual composition when made from anthracite as found by M. A. Adam's careful test on a 40 H.P. Dowson suction plant is given in Table 12, and also Mr Dowson's results from a similar series of tests on the same producer with gas coke as fuel in Table 12 *a**. The high heat efficiency of 90 °/ₒ recorded for the last six hours of the test when the plant had got settled down properly to its work, speaks well for the manner in which correct scientific principles have been embodied in the design of the Dowson producer. The composition of the gas may be taken as a typical analysis of the gas made in properly designed commercial suction producers, and we may say that with good fuel we have frequently

* *Producer Gas. Dowson and Larter, Appendix A.* (*Longmans.*)

TABLE 12.—*Trial of* 40 B.H.P. *Dowson Suction Plant with Anthracite.*

	Fuel					Gas					Heat efficiency	
	Calorific value by bomb calorimeter	Moisture	Ash and clinker	Quantity of fuel burnt	Total heat in fuel burnt	Analysis—mean composition	Calorific value		Total gas generated dry, at 0° C. and 760 mm.	Total heat in gas generated	Lower value	Higher value
							By calculation	By Junkers calorimeter				
	Brit. T.U. per lb.	Per cent.	Per cent.	lbs.	Brit. T.U.	Per cent. by volume	Brit. T.U. per cubic foot		Cubic feet	Brit. T.U.	Per cent.	Per cent.
Whole run—cold start	13,887	3·60	8·76	245·25	3,406,000	H 15·59 CH 1·31 CO 19·05 CO_2 5·93 O 0·79 N 57·33	132·6	—	21,890	2,903,000	79	8
Last six hours—hot start	13,887	3·60	8·76	180·00	2,500,000	H 15·64 CH_4 1·16 CO 20·13 CO_2 6·09 O 0·74 N 56·24	135·3	—	16,715	2,261,000	84	90

TABLE 12 *a.*—*Trial of* 40 B.H.P. *Dowson Suction Plant with Coke.*

	Fuel					Gas					Heat efficiency	
	Calorific value by bomb calorimeter	Moisture	Ash and clinker	Quantity of fuel burnt	Total heat in fuel burnt	Analysis—mean composition	Calorific value		Total gas generated dry, at 0° C. and 760 mm.	Total heat in gas generated		
							By calculation	By Junkers calorimeter			Lower value	Higher value
	Brit. T.U. per lb.	Per cent.	Per cent.	lbs.	Brit. T.U.	Per cent. by volume	Brit. T.U. per cubic foot		Cubic feet	Brit. T.U.	Per cent.	Per cent.
Whole run—cold start	12,477	—	11·37	276·2	3,446,000	H 12·0 CH_4 0·4 CO 25·5 CO_2 5·3 O 0·65 N 56·15	133·3	133·6	21,786	2,903,700	80	**84**
Last six hours—hot start	12,477	—	11·37	209·1	2,609,000	H 13·2 CH_4 0·35 CO 25·3 CO_2 5·4 O 0·6 N 55·15	136·2	136·9	17,053	2,322,500	84	**89**

obtained a gas from 150 to 160 B.Th.U.'s calorific value in National producers when the proportion of air and steam has been carefully adjusted.

Comparative Performances of Different Types of Producers. The usual commercial guarantees given with suction producers is that the consumption at full load will not exceed one pound of anthracite per B.H.P. and 1¼ lbs. of gas coke at full load in both cases. These consumptions are not intended to include stand-by losses due to shutting down the plant from stopping time at night until starting time the next day, and having regard to these and also the considerably fluctuating loads met with in practice, the all-round consumption, including stand-by losses, is nearer 1½ lbs. per B.H.P. on the average load over a working week. The only authoritative comparative tests ever carried out are those which were held by the Royal Agricultural Society of England at Derby in June 1906, when Capt. Sankey, M.I.C.E., and Professor Dalby, M.A., B.Sc., were the judges. The plants tested varied between 15 and 20 B.H.P. and the trials lasted over the whole of a week. The "National Plant" (Bickerton's and Robson's patents) was given the highest award on its design and performance, and a few of the tabulated results which appear in the official report are given below in Tables 13, 14 and 15. The full load test on anthracite included one night's stand-by losses, and the fuel was in every case measured most accurately by debiting each competitor with all the fuel supplied to him at the commencement of each test when the producers were cold and empty, and crediting him with the fuel returned at the end of the run, the producers being drawn immediately after the finish of each test and the hot fuel weighed after ash and clinker had been separated out. In computing the amount of fuel weighed back due regard was paid to its reduced calorific value on account of its having been partially burned or subjected to heat in the producer.

The usual method of testing the fuel consumption with suction plants is to start the trial with the fuel at a certain level in the generator and to finish back at the same level, all the fuel used meanwhile being weighed. By this manner of measurement, lower consumptions than those recorded at the Derby

DERBY TRIAL RESULTS (from Official Report)*.

TABLE 13. *Particulars of Full Load Trials with Anthracite.*

No. in Catalogue	Name of plant	Total hours running on full load. Banked 14 hours	Net coal used	Coal per hour of running time	Total water used	Water per hour	Average load on brake	Average revs. per min.	Brake constant	Average B.H.P.	Coal per B.H.P. per hour	Water per B.H.P. per hour	Per cent. of declared full load	Per cent. of Catalogue full load
			lbs.	lbs.	gallons	gallons					lbs.	gallons		
7	National	14·02	303	21·6	329	23·5	192·5	191·8	·000558	20·6	1·04	1·14	103	103
8	Railway and General	14·00	416	29·7	1012	72·3	220·0	170·5	·000533	20·0	1·48	3·61	100	100
9	Davey-Paxman	14·00	295	21·1	248	17·7	149·4	226·1	·000484	16·4	1·28	1·08	106	106
11	Dowson	14·08	312	22·2	264	20·0	189·4	192·4	·000559	20·4	1·09	0·98	102	102
12	Campbell..............	13·70	289	21·1	324	23·6	245·5	201·8	·000390	19·3	1·09	1·22	107	107
13	Campbell (Throttle).	14·00	336	24·0	393	29·6	290·6	188·8	·000389	21·4	1·12	1·38	107	107
16	Dudbridge	14·00	321	22·9	602	43·0	207·2	193·4	·000389	15·5	1·47	2·76	77·5	77·5
17	Morsey.................	13·93	307	22·0	668	48·3	280·0	203·5	·000343	19·5	1·13	2·47	97·5	97·5
19	Hindley	Trial	incom	plete										
20	Kynoch	14·00	297	21·2	180	13·4	115·2	240·0	·000668	18·4	1·15	0·73	108	108
21	Nowton	14·05			709	50·4	193·0	192·5	·000487	18·1		2·80		90
		Fuel re	turned	from	produc	er not	weighe	d						
28	Fielding	14·00	338	24·1	516	36·8	174·4	218·5	·000484	18·4	1·31	2·00	102	92
30	Crossley	Withdr	awn											
31	Crossley	14·00	247	17·6	517	37·6	195·3	178 7	000439	15·3	1·15	2·45	102	102

* *Report on the Trials of Suction Gas Plants. Royal Agricultural Society*, 1906. (*John Murray.*)

DERBY TRIAL RESULTS (*continued*).

TABLE 14. *Particulars of Full Load Trial with Coke.*

No. in Catalogue	Name of plant	Total hours run on full load	Net coke used	Coke per hour	Total water used	Water per hour	Average load on brake	Average revs. per min.	Brake constant	Average B.H.P.	Coke per B.H.P. per hour	Water per B.H.P. per hour	Per cent. of declared full load	Per cent. of Catalogue full load
			lbs.	lbs.	gallons	gallons	lbs.				lbs.	gallons		
7	National	9·00	224	24·8	258	28·7	191·9	190·8	·000558	20·5	1·21	1·40	102	102
8	Railway and General	9·00	255	28·3	710	78·9	207·5	183·1	·000533	20·2	1·40	3·90	101	101
9	Davey-Paxman	9·00	199	22·1	191	21·2	120·4	231·4	·000484	13·4	1·64	1·58	86·5	86·5
11	Dowson	9·00	232	25·8	320	35·6	193·3	188·8	·000559	20·4	1·26	1·75	102	102
12	Campbell	9·00	203	22·6	440	38·9	203·8	201·3	·000390	17·0	1·33	2·28	100	94·5
13	Campbell (Throttle)	9·00	214	23·7	330	36·7	231·3	190·5	·000389	17·1	1·39	2·12	95	85·5
16	Dudbridge	7·33	189	25·7	517	70·5	206·7	194·1	·000389	15·6	1·65	4·52	78	78
17	Mersey	9·00	213	23·7	875	97·2	240·0	205·5	·000343	16·9	1·40	5·70	84·5	84·5
19	Hindley													
20	Kynoch	9·00	222	24·7	284	31·6	112·3	240·4	·000668	18·0	1·37	1·75	106	106
21	Newton													
28	Fielding	9·00	250	27·8	440	38·9	176·4	221·0	·000484	18·8	1·48	2·06	104·5	94
30	Crossley													
31	Crossley	9·00	170	18·9	720	80	174·2	181·6	·000439	13·9	1·36	5·76	107	93

DERBY TRIAL RESULTS (*continued*).

TABLE 15. *Gauge Pressures measured from ordinary Indicator diagrams and from Mathot diagrams (full load only).*

No. in Catalogue	Name of plant	Maximum initial pressure, lb. per sq. in.				Compression pressure, lb. per sq. in.				Vacuum at beginning of compression stroke, lb. per sq. in.			
		Anthracite full load		Coke full load		Anthracite full load		Coke full load		Anthracite full load		Coke full load	
		Varied		Varied									
		from	to	from	to	Working stroke	Idle stroke	Working stroke	Idle stroke	Gas and air valves open	Air valve alone open	Gas and air valves open	Air valve alone open
7	National	240	390	260	340	160	150	155	140	0·2	0·5	0·5	1·5
8	Railway and General	220	380	190	340		95		90	0·9	4·5		
9	Davey-Paxman	200	310	150	230	115	85	110	75	1·2			
11	Dowson	300	390	250	360	140	130		120	0·3	0·5	0·5	
12	Campbell	340	420	360	420	155	140	150	120	1·0	3·0	1·2	2·0
13	Campbell (Throttle)	180	370	150	250	150		155		1·5		2·3	
16	Dudbridge	280	330			110	85			2·0	4·0		
		490	after	every	"miss"								
17	Mersey	300	360	170	270	120	110		90	0·5			
19	*Hindley	200	270										
20	Kynoch	240	390	190	345	145	130		120	1·8			
21	Newton	260	400			200	200			0·4	0·6		
28	Fielding	210	300	150	255	115	85		75	1·2	4·3		
30	Crossley	Not	indica	ted									
31	Crossley	435	510	360	490	210	190	195	165	0·5		0·5	3·5

* Mathot diagram only.

tests have been obtained, but the latter were unquestionably carried out on a more accurate principle than by the convenient method to which we have just alluded.

The anthracite fuel used at the Derby tests was carefully analysed and was found to contain 3·22 °/ₒ of moisture, and 6·44 °/ₒ of ash: the effective calorific value, determined by combustion with oxygen under pressure, was 7·475 calories per gramme, which correspond to 13455 B.Th.U.'s per pound.

Other recorded tests. We are indebted to the courtesy of Messrs Tangyes, Ltd., for particulars of a series of 10 hours tests carried out at their works by the well-known gas engine expert, Mr R. E. Mathot of Brussels, with the object of testing the fuel consumption and efficiency of their gas engines and producers. It is stated that the engines were taken from ordinary stock just as sent out for the customers and were not specially prepared or arranged for the purpose of the trial. It is to be noted that the fuel consumption was not tested by the same accurate method as that employed at Derby, and it must also be noted that there are no stand-by losses included, but the results are, nevertheless, very creditable. The consumption of fuel was ascertained as follows:

After cleaning the firebars and consolidating the fuel by a thorough poking, the generators were charged to a predetermined level. The weight of fuel added during the trials was noted at frequent intervals, and at the end of the trial the generator was refilled to the original level, the weight of fuel observed, then after poking and consolidating and cleaning the grate, the generator was filled up again and the total weight noted.

The engines were of the horizontal type, with a single cylinder, working on the four-stroke cycle with mechanically operated valves and magneto-electric ignition. The smaller engine, designated FS, was fitted with a single flywheel, and the larger ones with two flywheels outside the crank-shaft bearings. The engines FS and IS were governed on the "hit and miss" principle and JS by variable admission of the mixture by means of a throttle-valve controlled by the governor.

Rope brakes were arranged on the flywheels, one of the ends

being attached to the weight to be lifted, and the other to a spring balance, from which the necessary tension to regulate the friction was derived. Counting apparatus was fitted to record the number of revolutions as well as the number of explosions. Readings were taken at regular intervals.

TABLE 16.

Engine Reference Letter...	FS	IS	JS
Diameter of Piston ... . ins.	7	10	11
Stroke of Piston .... ins.	16	19	20
Revolutions per minute	220	190	180
Diameter of flywheels .... ins.	57	$63\frac{13}{16}$	$66\frac{1}{16}$
Thickness of rope:			
(1) For the positive load ....mm.	27	33	34
(2) For the negative load ....mm.	13	13	10
Radius to be taken into consideration:			
(1) For the positive load ....mm.	737	827	856
(2) For the negative load ....mm.	730	817	844

During the trials a series of diagrams and of graphics of explosions were taken, with a view to ascertaining the work indicated by the engines and the phases of the different cycles.

Calorimetric and chemical analyses of the gas produced by the generators were also taken, and in addition to this the quantity of CO in the products of combustion was noted. This did not exceed 1 %.

The coal was Scotch and Welsh anthracite of the following composition:

TABLE 17.

	Scotch	Welsh
Carbon	87·400	89·600
Hydrogen	2·600	3·250
Oxygen and Nitrogen ...	4·200	3·950
Ashes.......................	5·800	3·200
	100·000	100·000

The FS and IS were tested with Scotch anthracite.

The JS was tested with Welsh anthracite.

Remarks on the results of the tests. The dimensions of the engines FS and IS are amply sufficient for realising the powers of 8·5 and 20 brake-horse-power respectively. These powers are obtained with an average pressure not exceeding 65 pounds per square inch and the real average pressure attained 75 pounds per square inch.

The consumption of fuel is accounted for in two ways: first, by filling up the producers at the end of the tests before consolidating the fuel by poking the fire and clearing the grate; second, by filling up after consolidating and poking; the second being the more accurate method, and has been used to deduce the efficiencies.

The mechanical efficiency of the engines is very favourable, since it exceeds 80 °/ₒ, although the effective power remains about 15 °/ₒ below the figure which it may attain for the engines FS and IS. The thermal efficiency of 0·241, shewn by the engine JS, is most remarkable for an engine and producer plant of 29 brake-horse-power.

TABLE 18.

Performance.

Engine Reference Letter	FS		IS		JS
	Full load	Half load	Full load	Half load	Full load
Duration of test, hours	10	10	10	10	8
Average positive brake load, lbs.	103·86	51·43	231·70*	129·83	365·5
Average negative brake load, lbs.	17·97	3·96	34·7*	25·0	67·96
Average speed in revolutions per minute	225·06	226·61	197·06	197·58	186·05
Corresponding brake-horse-power	8·896	4·931	19·736	10·711	29·461
Average of the mean effective pressures, lbs. per sq. in. ...	73·816	71·242	74·811	72·962	81·638
Per cent. of useful cycles	85·8	58·6	84·9	54·3	
Indicated horse-power, based on the useful cycles.........	11·0933	7·4833	23·5516	14·73	36·34
Relation between brake and indicated power, per cent....	80·2	65·8	83·8	72·7	81·0

* +227·69 − 35·44 during two hours, and +232·69 − 34·51 during eight hours.

TABLE 19. *Fuel consumption and Heat distribution.*

Engine Reference Letter	FS		IS		JS
Operating condition	Full load	Half load	Full load	Half load	Full load
Gross hourly consumption of coal, measured before consolidating fuel by poking and cleaning fire, lbs..........	7·140	4·195	14·605	8·862	19·973
Gross hourly consumption of coal, measured after these operations, lbs.	7·738	4·744	15·505	9·987	20·910
B.Th.U. expended per hour (as per second method)............	109827	67335	220060	141742	312252
B.Th.U. in the coal per brake-horse-power	12345	13655	11150	13233	10598
B.Th.U. per brake-horse-power absorbed by cooling water of engine	3381	5030	3345	2840	3550
Per cent. of B.Th.U. represented—					
By effective work............	20·7	18·7	22·9	19·3	24·1
By loss through cooling water	27·7	37·3	30·4	21·7	33·9
By losses through the exhaust, radiation and the producer (by difference)	51·6	44·0	46·7	59·0	42·0
Total coal consumption per hour per brake-horse-power, by first method of measuring	0·802	0·85	0·74	0·827	0·678
Total coal consumption, etc., by second method	0·869	0·962	0·785	0·932	0·709
Residuals gathered under grates of generators after tests for the entire duration of same:					
Good coke, lbs................	1·5	2·37	2·0	1·99	1·7
Ashes, lbs.....................	3·36	4·5	4·0	2·14	3·7

Messrs Crossley Brothers publish results of tests carried out at Milford-on-Sea Electricity Works, where an engine developing about 40 B.H.P. consumed 0·89 lb. of anthracite per B.H.P. hour. In another case they instance an engine which developed 90 B.H.P. over a six hours run when the consumption of Manchester gas coke was 0·91 lb. per B.H.P. hour. In each case the fuel consumption was measured by starting and finishing with the fuel at a predetermined level in the producer.

Conclusions in foregoing results. The method of fuel measurement and the general supervision and conditions of the trials at Derby were such that the results obtained represent the very best that each maker was capable of doing with his particular plant, and these tests must, for the present at least, stand as the only really comparative and authoritative tests yet held. They shew, however, that the usual commercial guarantees of one pound of anthracite coal per B.H.P. hour without stand-by losses are justified by the best plants, but it is doubtful whether under the conditions of a really accurate trial starting from cold, this consumption can be improved on even with engines of 100 B.H.P. and over. In any case, the remarkable economy of combinations of engines and suction plants will be realised from the figures given, and in view of the established evidence which now exists of their general reliability and convenience, they constitute a great advance on the best steam engines.

Recorded results of behaviour under widely fluctuating loads. We have already referred in Chapter 9 to the difficulties which arise in working suction producers where the load rapidly fluctuates over a considerable range. The greatest variation which can possibly take place is, of course, that between 'no load' and 'full load,' or *vice versa.* The use of the Mathot explosion recorder enables us to study intelligently what takes place under these conditions, and Fig. 69 has been reproduced from consecutive explosion records taken from an engine working on a suction plant over a 15 minutes run while the change from 'no load' to 'full load' was taking place. These diagrams were taken from a 40 B.H.P. engine and plant, both being of the usual proportions for this working load.

Studying this diagram in detail it will be seen that during the greater part of the first minute's observation the engine was running on 'no load' and taking about one impulse in every five cycles. When the records were commenced the engine had been running without load for over an hour, so that the fire had got thoroughly settled down to the 'no load' conditions. It will be seen, however, that the gas was of sufficient quality to give good sharp explosions which compare favourably with those obtained whilst the engine was running on 'full load.'

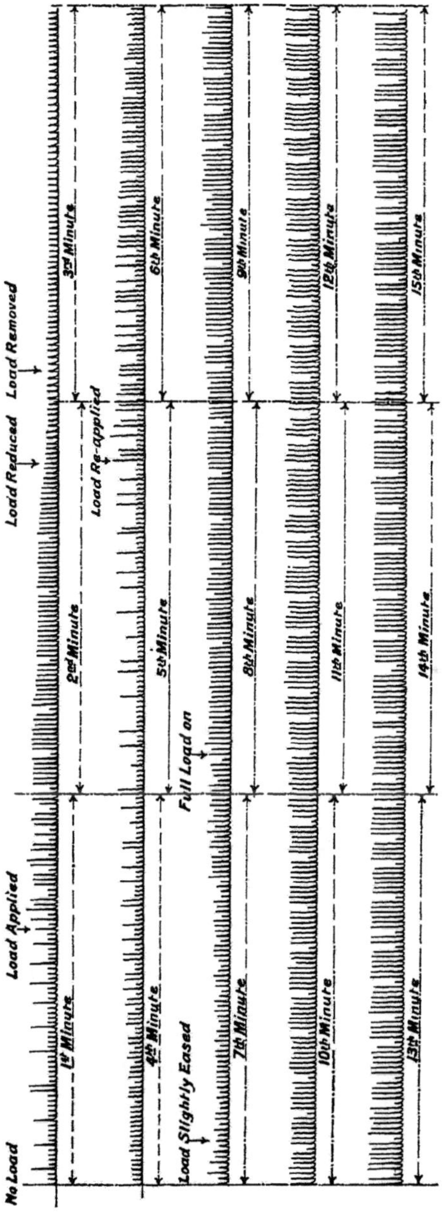

FIG. 69. Continuous explosion records taken from engine during change from 'no load' to 'full load.'

The load was applied at the point indicated towards the latter part of the first minute's record, and more frequent explosions are naturally indicated thereafter until at the commencement of the second minute the engine was taking a full impulse almost every cycle. It will be seen, however, that before the middle of the second minute the initial pressures commenced to fall until towards the end of this period they had almost died off altogether, so that the load had to be reduced: at the commencement of the third minute the charges had ceased to ignite in the cylinder and the load had to be removed. The engine was kept on turning due to the momentum of the flywheels, and the gas cock being kept open the engine continued to suck on the gas plant every cycle, with the consequence that at the early part of the fourth minute's record it will be noticed that the charges were again igniting, so that at the end of this minute a fairly good initial pressure was obtained. The load was re-applied towards the end of the fifth minute, and during the sixth minute the same falling off which took place during the second minute is observed, but in a less degree, seeing that on this occasion, although the explosions were weaker, the gas was always of sufficient quality to enable an ignition to be made. In order to keep the speed of the engine up, however, the load had to be slightly eased at the commencement of the seventh minute, but at the early part of the eighth minute it was re-applied, and by this time the fire in the generator had accommodated itself to the new conditions and the engine was able to carry its load thereafter, until at the end of the eleventh minute the explosion pressures were a maximum and were not afterwards increased.

It will, therefore, be seen that under the conditions described above seven and a half minutes were occupied in getting the plant on to 'full load,' and during this time the load had to be applied and removed twice.

The physical explanation of the foregoing is that, as pointed out in Chapter 3, the velocity of air passing through the fire has a determining influence on the quality of gas produced within certain limits. Now, whilst the engine is running on 'no load' and taking only an occasional explosion, the volume of pipes, scrubber, etc., which are interposed between the gas inlet to the engine and the air inlet to the furnace of the producer acts as a buffer and averages down the speed of the inflow from engine inlet to air inlet to the furnace of the gas plant, so that the latter is much less than the former. On the other hand, as so much less oxygen is being supplied to the gas plant under these conditions, the depth of fuel at a sufficient temperature to enable the necessary reactions to take place is naturally much less than when the engine is working on 'full load' and drawing in a correspondingly greater amount of air to the furnace.

If, therefore, we compare the conditions obtaining in the furnace when the engine is running steadily on 'no load' as against those when it is at 'full load,' we will find in the former a shallow fire through which the air to be reacted on passes comparatively slowly, thereby allowing the necessary reactions to take place in a sufficient degree to yield a gas which can easily be ignited in the cylinder of the engine. At 'full load' with the greatly increased air supply there will be a much greater depth of incandescent fuel, and consequently though the air passes through at a much higher velocity it is in contact with the incandescent carbon for a sufficient time to enable the reactions to take place. When a sudden change occurs, however, from 'no load' to 'full load,' an intermediate state between the two just described must evidently be passed through, for, when the load is suddenly applied, the buffer action of the scrubber, etc. volume is felt in a much less degree, and a great increase in the velocity of air inflow through the furnace is immediately set up through the engine taking a full charge of gas every cycle. At first the effect of this is merely to empty the scrubber of the burnable gas contained therein from the previous conditions of running, and consequently for about a minute the charges continue to be ignited in the cylinder. At the furnace, however, the suddenly increased inflow of air, both in velocity and amount, is immediately felt, and this air passing quickly through the shallow fire (which we have pointed out is a natural consequence of the 'no load' condition of running) does not react with the fuel in such a way as to produce a burnable gas. It merely causes a more rapid combustion to be set up in the shallow fire and a gas richer in carbon dioxide (CO_2) results. The hot zone, however, under these conditions rapidly enlarges until, as we have seen from Fig. 69, at the end of the eighth minute it is in such a condition as to meet the 'full load' requirements.

Mr Dowson has analysed the gas made during the transition period and found it to be under 100 B.Th.U.'s per cubic foot, and in most cases it will be found to be even below this: in any case it is usually of too low a calorific value to admit of ignition in the engine cylinder.

It will be realised from the foregoing that any attempt to automatically control the steam supply will not assist the matter at all, for the fire in the producer will die down under 'no load' conditions due to the decreased supply of oxygen, whether the steam be automatically regulated or not. A reduced fire must always be enlarged for 'full load' conditions, and during the intermediate period the engine must suck out the bad gas made: missed ignitions will result as shewn in Fig. 69, and as the speed of the engine consequently tends to fall the governor will come into action and automatically cause the gas valve to be opened to the full extent, whatever be the system of engine governing adopted. Any arrangement of automatic water or steam regulation depending upon the degree of engine suction (which is the usual plan adopted) would, therefore, be of no avail in helping the plant along at the critical period. On the contrary, it would call into play an increased water supply just at the particular time when, if anything, it would be desirable to restrict the same so that the temperature of the fire might be allowed to rise as quickly as possible.

The difficulties mentioned above are easily got over by properly arranging installations to meet the special requirements, if any, which exist and, although at first sight it might appear to be a drawback that an engine working on a suction plant cannot immediately respond to a great increase of load, it has not been found in practice to be a difficulty affecting the successful application of suction gas power in all cases. When extreme changes of load have to be met with it is usual to fit a bye-pass arrangement from gas main into engine cylinder so that a certain quantity of gas is drawn into the cylinder of the engine every cycle whether the governor allows the gas valve to be opened or not. The quantity of gas drawn in from this independent supply is not sufficient to ignite in the cylinder, and, therefore, it does not in any way interfere with the speed regulation of the engine, but by this means a steady and continuous suction action upon the fire is maintained during 'light load' conditions, and though a certain amount of gas is thereby wasted, the loss of economy may for all practical purposes be considered negligible. As a result of this steady suction, the fire is kept in good condition to meet

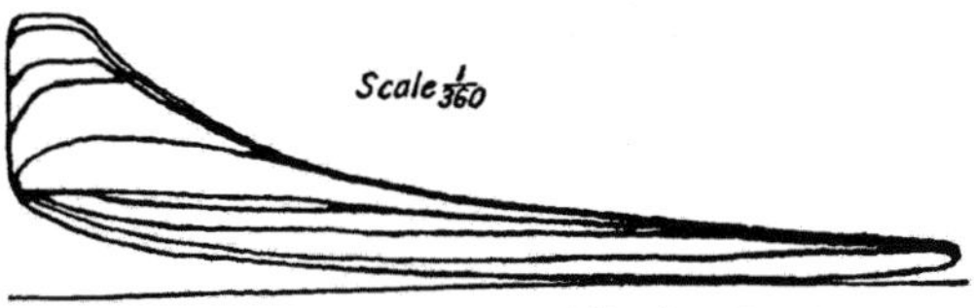

FIG. 70. Indicator diagrams taken while changing from 'no load' to 'full load.'

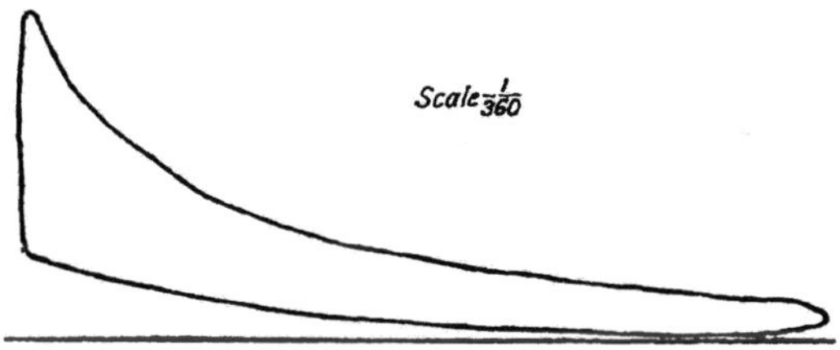

FIG. 71. Normal power indicator diagram in 7½ minutes after load applied.

the increased load requirements when the engine is called upon to suddenly respond to the same.

No difficulty is experienced when a change of load in the other direction, namely from 'full load' to 'no load,' takes place, as obviously under such conditions the fire will gradually die down but a burnable gas will always be produced.

We have purposely considered, in the course of these remarks, the very worst conditions under which a suction plant may be put to work. It is seldom, however, that these great extremes occur in practice, and under the ordinary fluctuations of load, say from 'half load' to 'full load,' there should be no period of missed ignitions at all, and the gas plant should respond to any charge within this range without the slightest difficulty. Fig. 70 shews a series of ordinary indicator diagrams taken with the pencil of the indicator held to the drum while the load was being applied during the first two minutes of the continuous explosion records shewn in Fig. 69.

Fig. 71 shews the normal power indicator diagram taken at the middle of the eighth minute (see Fig. 69).

SECTION II.

PRESSURE PLANTS FOR NON-BITUMINOUS FUEL.

CHAPTER 11.

GENERAL PRINCIPLES AND DESIGN.

WE have seen from the previous chapters that the fundamental idea of the Suction Plant is to cause the engine to exert a suction effect direct on the gas producer, whereby the make of gas is automatically regulated by the engine itself to suit the load it has to meet. The alternative method of producing gas and supplying it to the engine differs from this system chiefly in so far as an air and steam blast at a pressure of about 10 inches water gauge is forced through the producer and, with this blast behind it, the gas produced at the plant and supplied through the mains ready for use, is at a pressure of about 3 inches water gauge. It has hitherto been found necessary to work all bituminous fuel producers with a positive blast of from 15 to 24 inches pressure in this way, and for many purposes, even when anthracite fuel or gas coke is to be used, it will be found distinctly advantageous to use the blast in preference to the automatic suction effect of the engine. When the producers are supplied with a positive blast at pressure in the way described the plants are designated *Pressure Plants,* as distinguished from Suction Plants of the types we have already described. The present section deals with pressure plants using such non-bituminous fuels as those referred to.

It has already been pointed out at the conclusion of Chapter 2 that one of the chief features which has marked the introduction of the suction plant is the general recognition of the fact that the sensible heat of the outgoing gases leaving the producer can be utilised for the purpose of raising the necessary steam which is required for use in the producer itself. Hitherto in the pressure plants, made by Messrs Dowson, Tangyes and other firms, it had been customary to have a separate steam boiler to generate the steam required for the producer. When this plan is adopted the general reactions in the producer, and the heat balance as stated in Tables 4 to 8 in Chapter 2, remain practically the same, with the exception that the heat necessary for steam raising purposes need not be deducted from the left hand of the equation in all cases. This would seem to point to further heat being available for the endothermic reactions of the steam on incandescent fuel. As already pointed out however, the gases leaving the producer are still at a temperature of over 600° F. when they have raised the necessary steam in a suction plant, so that even thereafter they contain quite sufficient heat for steam raising purposes. In a pressure plant with an external steam raising boiler, the gases leave the producer at quite as high a temperature as in a suction plant if the latter had no vapouriser or preheating device, and therefore we may say that the pressure plant with independent boiler is less efficient than the suction plant in so far as, firstly, there is less recovery of the waste heat in the producer, and secondly, additional fuel is required for firing the independent boiler.

To get over this difficulty Messrs Crossley Bros. introduced a pressure plant of the general type shown in Fig. 72. The producer is a cylindrical firebrick lined chamber, the fuel being introduced by means of two feeding hoppers on the top, to be used alternatively at one hour or more. The air and steam pass upwards through the fuel and become converted into gas, which is taken away by a central collecting bell at the top, which also defines the depth of the active fuel and always keeps this constant so long as the fuel is kept above the lower mouth of the central collector. It will be seen that at the bottom of the producer there is a water lute to enable the ashes and clinker to be withdrawn without interfering with continuous working, and to further facilitate the latter, there are inspection doors, and means for rotating the grate, which saves poking. After leaving the producer the hot gas is conducted through the saturator,

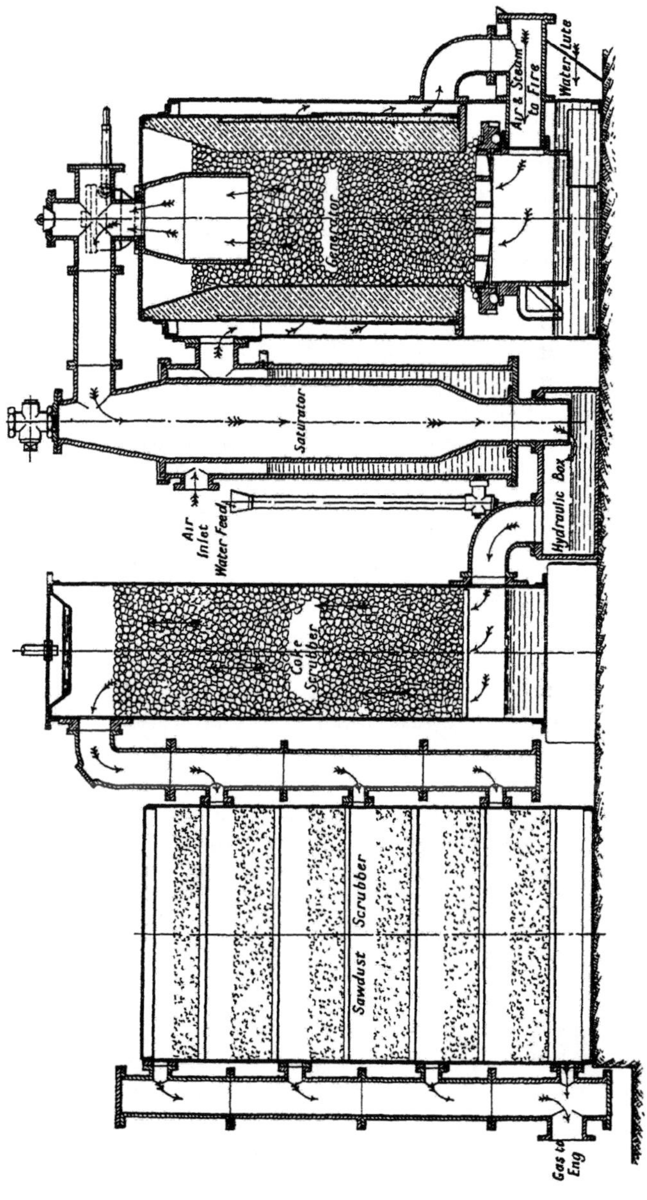

FIG 72. Sectional diagram of Crossley pressure plant for anthracite or coke.

which is a concentric tubular arrangement, the hot gas passing down the central pipe, and water is placed in the annular space between this inner pipe and the outer shell, and is vapourised by the heat of the former. Air supplied by a power fan is blown into the saturator and catches steam which is made there, and the mixed air and steam are superheated by being passed round a jacket which encircles the whole of the upper part of the producer. The heated air and steam are then taken underneath the grate and passed through the fire, enabling the required reactions to take place. There is a coke scrubber and a sawdust scrubber with sections through which the gases pass in parallel. After leaving the sawdust scrubber the gases are either taken direct to the engine or they may be passed into a gas holder, for as the air is being forced into the producer by a fan, if at any time a back pressure is produced through the gas not going forward, on account of the requirements being small, it simply means that the fan churns the air, for it is a feature with fans that they cannot work against anything but a very small pressure.

The action of the producer which we have just described is identical with the suction plant, except that instead of the engine drawing forward its own air, a power driven fan is used for the purpose. In such a manner any ordinary suction plant can be converted into a pressure plant by fitting a fan in a similar way to that just described. Messrs Crossley Bros. claim an efficiency of 88 °/$_\circ$ for their pressure plants on these lines, though it is not quite clear whether this figure includes the power required for driving the fan. With the latter exception there is no reason why in a plant of this sort the efficiency should not be as high as in an ordinary suction producer, which we have seen may be as much as 90 °/$_\circ$.

To Mr J. Emerson Dowson of London, however, belongs the credit of introducing the first plant to work on the principle of making a gas suitable for engine work by forcing the mixture of air and steam through the producer, *and his apparatus was the first successful power gas producer made.* The general form of the same is shewn in Fig. 74 and a section through a typical Dowson generator is shewn in Fig. 73. Steam up to a pressure of about 60 lbs. per sq. inch is raised in a small subsidiary boiler, and this steam at pressure is used to force the necessary air into the producer by the simple form of injector shewn in Fig. 78. The gas after leaving the producer is passed through a hydraulic seal, then through the coke scrubbers and sawdust scrubbers to the gas holder. The effect of the whole is to put at least 3″ of pressure on the gas main, and this makes the hydraulic

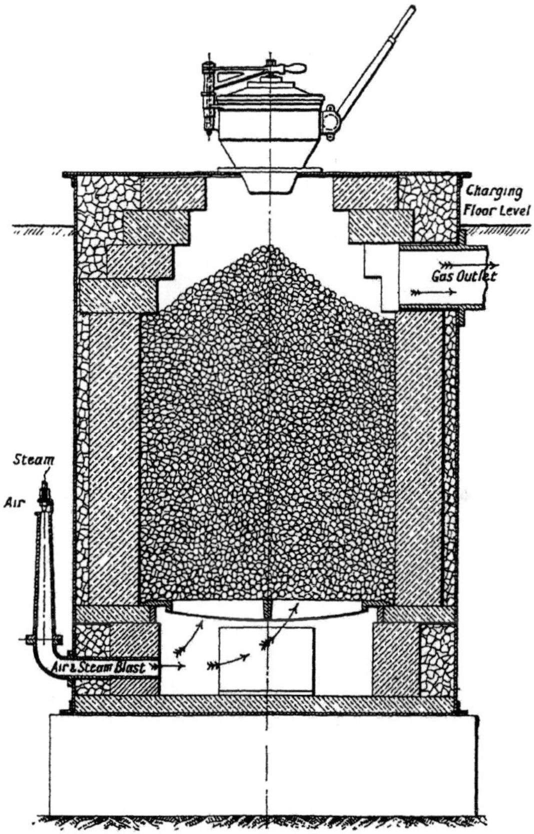

FIG. 73. Section through Dowson pressure generator.

box which prevents the gas coming back through the scrubbers to the producer of much more importance than in the case of the suction plant. Although there is not the same degree of regeneration in these plants as in the suction plants and the Crossley pressure plant which we have just described, it will be at once seen that the independent boiler is not wholly a disadvantage, for it makes the plant much more self-contained than when a power driven fan is used. The latter requires to be driven by a gas engine, steam engine, or other prime mover: the gas engine implies the existence of an independent gas supply; an oil engine, if employed, requires considerable time to start up and also continual attention, whilst an electric motor requires an independent electric supply. At the sacrifice of a slight loss of efficiency, the steam boiler gets rid of all these contingencies, and in starting up the plant the attendant merely gets steam

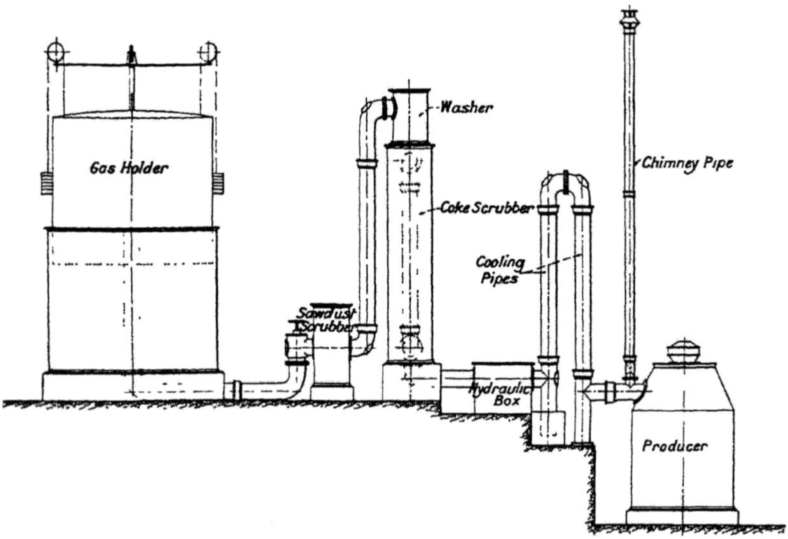

FIG. 74. General arrangement of Dowson pressure plant.

pressure on his boiler, and he is then ready to blow up the fire in his producer, and gas is made very quickly and conveniently. It is also unquestionably a great advantage to have a gas holder, especially where heating work is being done as well as where gas for power purposes is required: for in pressure plants as in suction plants it is more difficult to make gas of the right quality when the plant is running at light load than when at full load. Mr Dowson's arrangement, however, satisfactorily meets the conditions of light load with his pressure plant, for it is possible to fill up the gas holder intermittently. For instance, when the attendant sees the gas holder getting low, he turns on his steam and blows up the fire, and allows the poor gas so made to blow away to the chimney. A steady blow of a few minutes brings the fire into a first class condition suitable for making a rich gas which he forces into the holder until the latter is quite full again, he then stops the steam and air blow until the holder begins to get low again. It will be seen that by exercising a little judgment in this way it is possible to prevent any bad gas being sent to the heating appliances or to the engine, no matter what great fluctuations of load there may be. We regard this as being of special importance where heating work has to be done, for as we have already indicated in Chapter 9 it is essential that there should be no risk of the gas jets in any heating appliance becoming extinguished through the quality or pressure of the gas falling off unduly. The introduction of a gas holder and the working of a pressure plant on the lines we have just described constitutes in our view the safest arrangement for heating work, or for a combination of heating and power work. So far as any slight gain of efficiency or reduced first cost is concerned, which may be conferred by the adoption of regenerative plants without holders, we say unhesitatingly that whatever the advantages gained in any such way, such advantages are of small account compared with the safety of the workpeople using the gas, and from this point of view it is vital that proper means be adopted to absolutely ensure a steady supply of gas at a uniform pressure and heating value, and in our opinion there can be no question that so far as plants using non-bituminous coal are concerned, the Dowson type of pressure plant, combined with a gas holder, is the safest and best arrangement.

Messrs Tangyes have adopted the Dowson principle in their pressure plants, a typical one of which is illustrated in Figs. 75 and 76, the only difference being that the outlet pipe from the generator is passed through a water jacket to assist in the cooling, and the steam supply from the boiler to the air injector is passed through a superheating coil placed in the gas outlet pipe immediately adjacent to the producer.

The National pressure producer shewn in Fig. 77 is generally similar to the Dowson type but embodies many constructional

FIG. 75. Tangye pressure plant.

merits to ensure continued efficiency with minimum upkeep. Cast iron is introduced where possible to obviate excessive corrosion; the ashpit and furnace doors are machined and fitted on the face and provided with stout asbestos packing so as to make a good joint without the use of fireclay luting, and as the fuel used is most frequently gas coke which forms a good deal of clinker, poking holes are provided in the top of the producer to dislodge the same from the sides of the furnace.

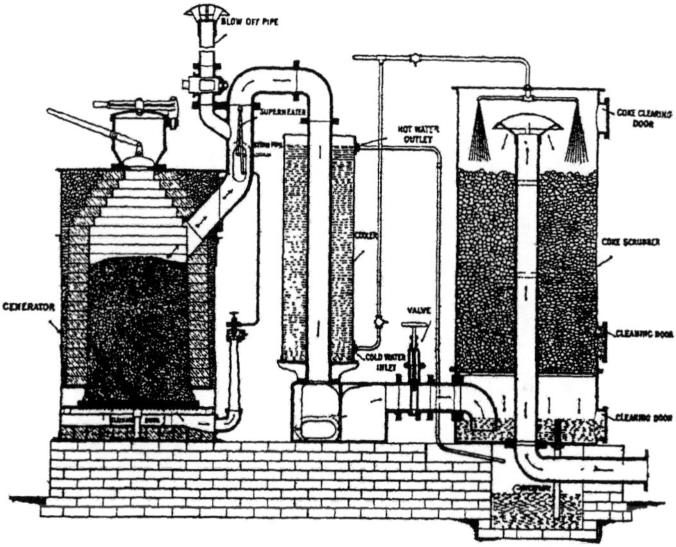

FIG. 76. Section through Tangye pressure producer.

Steam and air injectors. The general principle on which injectors work is well understood and the injector appliance used for inducing a pressure of air under the grate of the producer is similar in principle and action to the feed injector on a steam boiler. Three forms of steam and air injectors are shewn in Figs. 78, 79, and 80; the actions in all being the same. Dry steam issuing at a high velocity from the jet rapidly expands in the blast pipe while still maintaining considerable velocity. The result is that the air in this pipe is driven forward and a partial vacuum is formed at the air inlet through which a constant

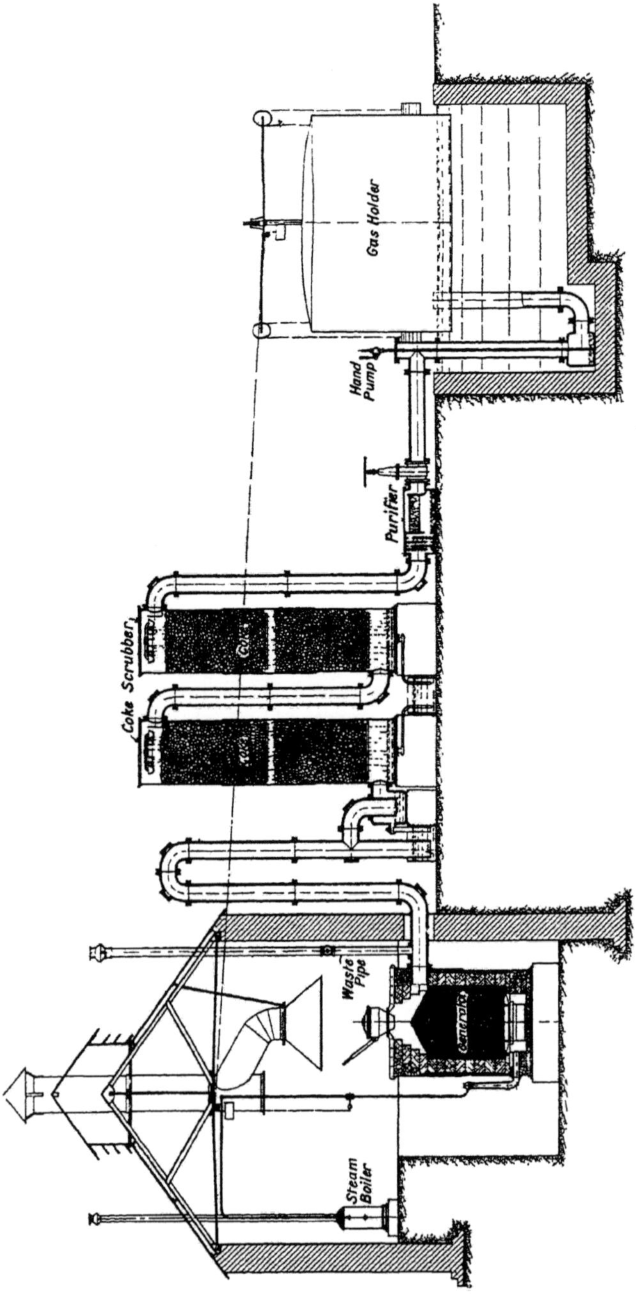

FIG. 77. Sectional elevation of National pressure plant.

stream of air consequently passes. The velocity head of the air and steam of the upper part of the blast pipe is transformed into pressure head at the point of entering the ashpit by gradually enlarging the sectional area of this pipe as shewn in the figures.

It will be understood from the remarks already made that these injectors should be sufficiently efficient, that when the proper amount of steam required for the best producer efficiency

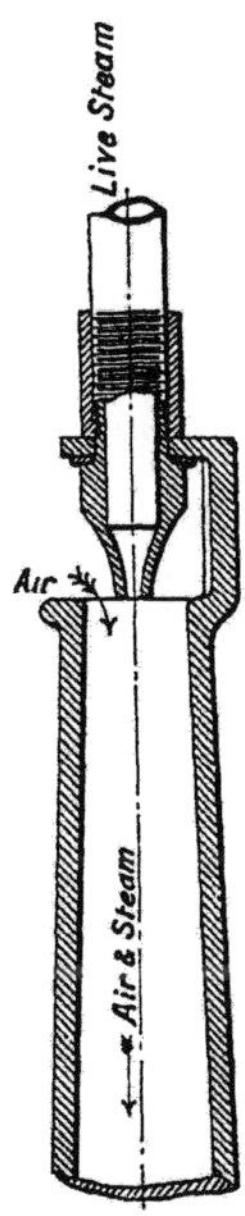

FIG. 78. Simple steam and air injector (Dowson).

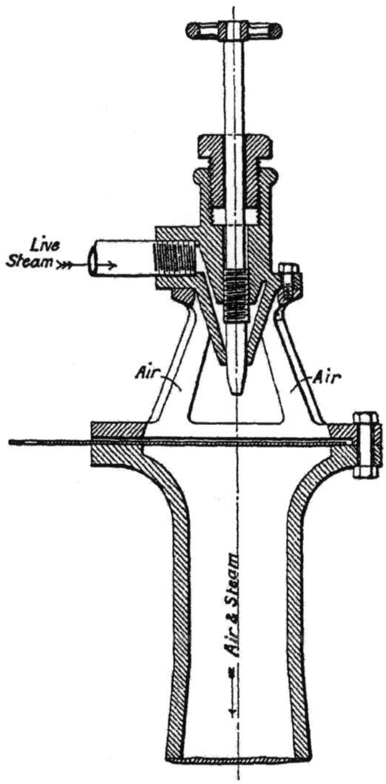

FIG. 79. Combined nozzle with steam regulator.

is issuing from the jet, the correct amount of air should also pass in. It is naturally somewhat difficult to ensure that these correct proportions will be obtained with any injector, and we consider the best plan is to arrange that the proper discharge of steam will drive *more* air forward than is actually required, and then to have means of throttling the latter as desired until the best results are secured. Besides, some fuels tend to form more

11—2

clinker than others, so that in the same producer it is frequently found desirable to alter the proportion of steam and air to give the best all round results. The design in Fig. 80 is particularly elastic from this point of view; the concentric steam nozzle with inside and outside air supply secures a high efficiency at the jet and the adjustment of both air and steam is provided for by a simple arrangement.

A $\frac{3}{4}$ inch diameter steam supply and $2\frac{1}{2}$ inch air inlet at the jet are sufficient for a 300 H.P. producer injector which if

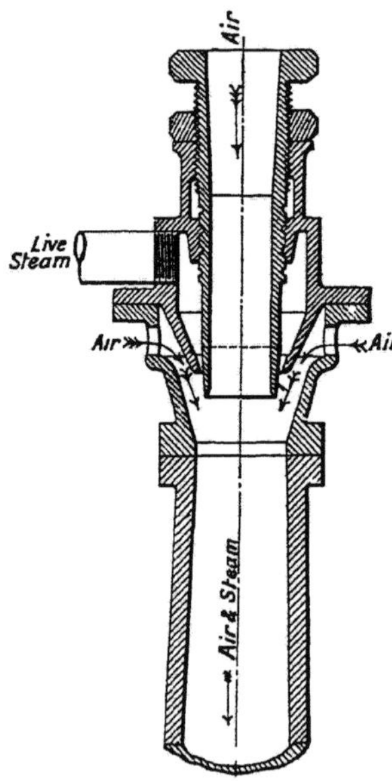

FIG. 80. Improved injector with both air and steam regulator (National).

designed on these general proportions will be found to give good results.

Fans or blowers. If fans or blowers be applied to create the air pressure instead of the injector just described, their proportion can be fixed on the basis of a delivery equal to 100 cub. ft. of free air per pound of fuel to be consumed per hour. This is a great deal more than is theoretically required (see Chapter 2), but in starting and at certain other times it will be found convenient to have a comparatively powerful fan.

Steam boiler. Any good make of small vertical steam boiler will serve the purpose and the working pressure should be 60 lbs. per sq. inch. As regards size, we saw in Chapter 2 that the maximum amount of steam which can be decomposed per pound of carbon reacted on is 1¼ lbs. On the basis of 2 lbs. of steam per pound of carbon so as to get a boiler sufficiently large that it can be fired very easily, we would have for a 300 H.P. producer a boiler capable of evaporating 600 pounds of steam per hour. Reckoning that the power of small steam boilers is usually computed on the basis of 30 lbs. of steam per hour, a boiler for a 300 H.P. gas plant would then be styled commercially as a 20 H.P. *steam boiler.*

These boilers are usually fired with coke, and together with all pipes connecting the steam supply with the injector at the producer they should be well-coated with good non-conducting material of at least 2 inches in thickness. This covering is important, as it is essential to the good working of the producer that dry steam be supplied.

If a steam supply is already available where the gas plant is being installed, an independent boiler will not be required, but the steam supplied to the plant must be dried by passing through a suitable separator or superheater before it is used.

Grate and furnace area. The general proportions of the interiors of these producers may be similar to the suction producers described in the previous chapters, but as the make of gas is not intermittent in the case of pressure plants, the furnaces of the latter need not be so large power for power. A furnace area of 6 sq. inches per H.P. will be found to give good results; the depth of fire may be as in the suction type of plant.

Scrubbers and coolers. These are similar to those already considered in the previous chapters and usually consist of one or two cylindrical coke scrubbers according to the size of the plant and a final sawdust scrubber. When the sensible heat of the gases leaving the producer are not recovered for steam raising purposes rather more cooling will require to be done in the scrubbers and consequently more cooling water will be consumed.

Hydraulic box. When the gas is discharged into a holder which maintains a constant pressure in the gas mains, it is clearly important to have a good hydraulic box or water seal between the producer and the pressure so that there is no return of the gas to the producer between blows or at any other time. It is usual to arrange this seal immediately after the producer as in Fig. 77.

Gas holders. The construction of these small gas holders requires little detailed description. They are built of light mild steel plates and are of from 10 to 20 feet in diameter for powers up to 500 B.H.P. It is an obvious advantage to have them as large as circumstances permit. The tank for the holder can be made of brick built into the ground, or of mild steel fixed on or above the ground level.

Fuel used. The fuels suitable for use in these plants are anthracite, gas coke and charcoal. Anthracite peas are generally used on account of their comparative cheapness; the gas coke should be well carbonised and free from sulphur, and should not form excessive clinker; the charcoal and coke should both be broken in small pieces before use so as to obtain a reasonably dense fire.

Working results. In practice it is usual to guarantee a consumption of one pound of anthracite per B.H.P. per hour and 1¼ lbs. of well carbonised gas coke per B.H.P. hour with plants of 50 H.P. and upwards excluding stand-by losses and the fuel required for firing the boiler; these results can be obtained when the producer is working at full load. On light loads, however, a considerable quantity of gas has to be blown to waste through the quality falling off, and the efficiency of the plant is very considerably reduced thereby, but down to ¾ load there is no necessity to blow gas to waste.

Mr Dowson considers that, generally speaking, gas of a higher calorific value can be made with his pressure plants as against suction plants, and he gives the figures as being 161 B.Th.U. per cubic foot for pressure plants as against about 135 B.Th.U. for suction plants. With the ordinary degree of care however, which is accorded by the average attendant to combined gas plants and engines, we are of the opinion that the calorific value of gas made in suction plants is usually as high as that made in pressure plants, and speaking from a wide experience we have been able to obtain quite as much power from engines working from suction plants, when the latter have been reasonably near the engines, as we have found possible when the gas was supplied by pressure plants. This has specially been the case with smaller size engines from 10 to 30 B.H.P. and may be partly due to the increasing care which has lately been taken to adapt the engines to work on producer gas, but we also base this opinion on careful estimates of the calorific value of both suction and pressure gas taken over many readings with Junker's Calorimeter. Further, if the steam and air proportions are correctly adjusted in both cases there would appear to be no reason why, with the same fuel, the composition of the gas made should vary in the two cases. In many types of suction plants, however, the proportions of steam and air cannot be adjusted at all, whilst in pressure plants the proportion is precariously settled by the design of the injector, which may not always be correct, and with the steam jet type of pressure plant generally we consider that more efficient air injectors are required than those usually applied. In the latter, the importance of securing the proper proportions of steam in the air supply has been fully emphasised in Chapters 2 and 3 and the proportions there laid down apply equally to the type of plant now under consideration. It seems to us, therefore, that the air injector should not have its air supply regulated entirely by the steam but it should be possible to regulate both air and steam independently so as to secure the correct proportions under all conditions of fuel and load, and this has been secured in the National Pressure Plants with gratifying results.

CHAPTER 12.

CONSTRUCTION, MANAGEMENT, APPLICATION TO PRACTICE.

Construction of producers. The elimination of a vapouriser from the producer enables the latter to be made of an extremely simple form, and in the Dowson, National and Tangye pressure plants the producer is simply a cylindrical shell made of mild steel plate with an ashpit and grate in the bottom and a good fine brick lining around the side of the furnace. It is most important that this fire brick lining should be a thoroughly good job and the principles laid down in Chapter 3 with reference to suction plants apply equally here. It is specially important that the joints of the lower layers of brickwork be made carefully, as otherwise the air blast, which is at pressure, will force its way through and thence up the cavity between the bricks and the shell and combustion will be set up at the upper part of the producer with ruinous results.

In view of its simplicity of form, a steel plate shell is the usual practice in this type of plant; the cast iron fire and ash-pit door frames being riveted or bolted on. As it is necessary to renew the brickwork from time to time the top plate of the producer should be easily removable. This plate should preferably be made of cast iron as steel plates buckle.

In the Crossley producer the addition of the superheating jacket to the producer for the purpose of preheating the air and steam blast, effects a certain saving of heat which would otherwise be lost, but in view of the established performances of plants without this device, the practical advantages of it are doubtful. Moreover, steel plates rapidly corrode in the presence of hot air charged with moisture and renewals are most difficult. Generally speaking, no departure from simplicity of construction should be considered without a definite and sufficient gain in performance.

Automatic control of steam supply. When a holder is used in conjunction with a plant in which the air pressure is obtained by means of a steam injector, it is obviously a simple matter to arrange that the holder when full will control the steam supply to the injector, thus automatically regulating the make of gas. When the plant is running on very light loads, however, and a considerable time elapses between the holder being full and empty, this automatic control is not sufficient and the attendant must see that the gas is sufficiently good after each pause before sending it forward to the holder. It is usual in such cases to open the hopper valve to burn the bad gas away before allowing it to pass on to the holder.

Scrubbers. These are cylindrical in form and there is a great advantage in their being as large as possible. Their capacity should be at least 0·75 c. ft. per B.H.P. and the riveting and caulking work in connection with the steel plates should be carefully carried out. The sawdust scrubbers should be of ample size and arranged so that the sawdust can be easily taken out and the apparatus cleaned.

In connection with the water supply to the scrubbers and other details, the particulars which apply generally here will be found in Chapter 5.

General arrangement of plant. While the plant is at work the attendant's whole duty is to keep steam in the boiler and to feed fuel into the top of the producer. It has therefore become usual to place the latter in a pit so that the top of the producer is approximately at floor level (see Figs. 75 and 77). The pit must be of sufficient size to allow the clinkering bar to be worked on both sides of the producer and for easy access to all fittings. The scrubbers and gas holder can be arranged in any convenient relation to the producer to suit local conditions on site, but accessibility to all cleaning doors must be carefully borne in mind.

All gas connecting pipes about the plant should have cleaning doors fitted, and drain syphons must be fitted where necessary for draining off condensed moisture which may accumulate at the lowest part of each set of pipes and which if allowed to remain will prevent the gas passing through. A large hood and chimney pipe must be fitted over the hopper valve, so that when air gas is being burned to waste there is a free outlet for the waste gases.

Working and Management.

To start the plant. (1) Steam must first be raised in the boiler and starting from cold this will occupy about three-quarters of an hour.

(2) One of the furnace doors of the producer having been removed or opened and the hopper valve and chimney cock opened, a starting fire of wood, oily waste and shavings is set going. and a quantity of coke dropped on through the feeding hopper. When it is seen that these are burning well, further

coke or anthracite is added gradually until a depth of several inches of this fuel is seen to be well alight.

(3) The furnace doors may now be closed, likewise the hopper valve, and a little steam turned on to the injector; this will cause a draught up through the fire and out at the chimney cock. Further fuel must now be fed in through the hopper valve until a depth of fire equal to about 2 feet is obtained, the blast being increased meanwhile by turning on more steam to the injector. In from 30 to 40 minutes after first lighting the fire, a red hot bed of fuel will be obtained in the producer sufficient to yield a burnable gas at the test cock provided near the chimney pipe and it is usual to continue to waste the gas by burning it at the mouth of the hopper until it shews an orange tinted flame, which indicates a gas sufficiently rich that it may be turned into the holder ready for use.

(4) Regarding the depth of fuel to be maintained in the producer while the plant is at work, it should be realised that with gas coke or anthracite the furnace should be kept fully charged up to the level of the gas outlet pipe. There is no advantage but rather danger of making weak gas in working with a shallow fire.

(5) When the holder is being charged for the first time it must be remembered that prior to the gas being forced in, the holder and scrubbers are full of air which must be displaced. While the displacement is taking place there is a mixture of gas and air present, which during one period of the charging will be highly explosive, and consequently if there be any open jets in the proximity of the plant to which a light is inadvertently applied, a violent explosion may occur with destruction of the holder or some other part of the plant. Hence there is usually an outlet pipe from the holder to atmosphere, which is kept open while the first charge of gas is being forced into the holder and through which the displaced air discharges as the gas comes in. It is well to keep this outlet open for at least five minutes and any test cocks on this side of the plant or gas mains should be fitted with gauge mantles after the style of safety lamps to preclude any possibility of an explosion.

Working the producer. (6) As regards working the producer, some poking will be required when gas coke is used, but an anthracite fire should not be touched while the producer is in use. In all cases it is well to loosen during meal hours any clinker which may have formed on the grate, and of course this should be done carefully before the commencement of each day's run. Where continuous working night and day is an absolute necessity it is best to have two producers which can be used alternately. In any case the length of time which a producer may be worked without withdrawing the fire depends very largely upon the attendant's judgment and attention to duty in the systematic removal of the clinker in the manner described.

(7) When it is desired to shut the plant down for a short stoppage, it is usual to nearly shut off the steam to the injector, open the hopper valve and to light the gas issuing therefrom, which burns to waste for the time being. This keeps the fire bright and the plant is ready for gas-making again in a few minutes. When stopping for the day the vent provided by the chimney pipe is sufficient to induce a draught through the fire which is consequently kept alight. In this case the air enters at the injector pipe and to prevent waste of fuel furnace doors and hopper valve are kept shut.

Regular cleaning and attention. (8) When a producer is running at its full duty the fire should be withdrawn every week and all clinker removed from the furnace, care being taken to do no damage to the brickwork meanwhile. It is not good to draw the fire while the producer is very hot immediately after it has been in use, as the sudden chill which is set up contracts the heated brickwork too quickly, with the result that it cracks.

(9) All water seals should be carefully cleaned every week and deposits of dust removed from them.

(10) The coke scrubbers require similar attention to those of suction plants (see Chapter 8, p. 113). The sawdust scrubber requires attention every 3 to 4 weeks; the dirty sawdust should be replaced with new material, the whole being

well soaked before being put in. Generally speaking regular attention is essential to success.

(11) Before cleaning out scrubbers or holders, air and steam should be blown through the generator, the fire having been withdrawn: the scrubbers etc., thus may be cleared of the gas left in them from the working of the plant. This operation must not be hurried, and it is well to fill and empty the holder four or five times before attempting to enter it.

(12) The steam boiler should be regularly examined by a competent boiler inspector and in working the water gauges and other fittings should be left in perfect order. The water should be blown out every two weeks and more frequently if of a sedimentary character.

Application to practice. There is no doubt that for power work suction plants have very largely superseded pressure plants and the application of the latter is now almost entirely confined to cases where the gas is required both for heating and for power in which event the considerations stated in the previous chapter naturally incline the user to adopt them. In some few other special types of installations with very wide fluctuations of load pressure plants are desirable as the lag of the fire in suction plants causes the engine which they are driving to lose speed if the load is too suddenly applied after running at light loads for some time (see Chapter 10, p. 146).

It is further our view that plants of the pressure type just described can be economically applied up to 1000 H.P. if a cheap supply of gas coke can be obtained, in which case they work with less attention and general trouble than bituminous plants. We refer here to a central producer plant from which the gas made is distributed to various points within a reasonable distance for use with motors and for heating. Fig. 81 shews a 900 H.P. plant consisting of one Dowson and two National producers arranged for supplying a large engineering works, which is working with most satisfactory economy.

Fig. 82 shews a 250 H.P. central station installation running continuously night and day without stoppage. A regular consumption of 1 lb. of anthracite peas per B.H.P. hour has been shewn here.

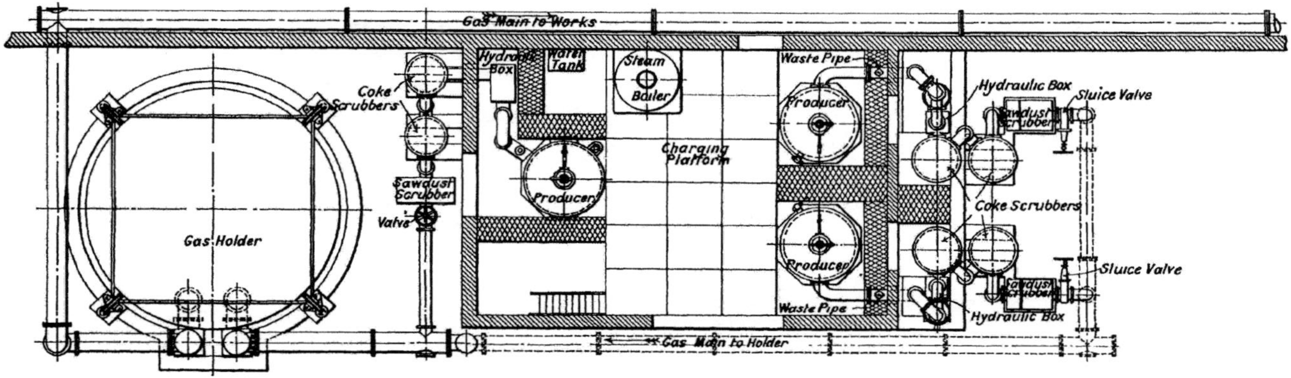

Fig. 81. 900 B.H.P. pressure gas producing plant.

There are many thousands of horse power of pressure plants such as those described at work successfully, and they are being

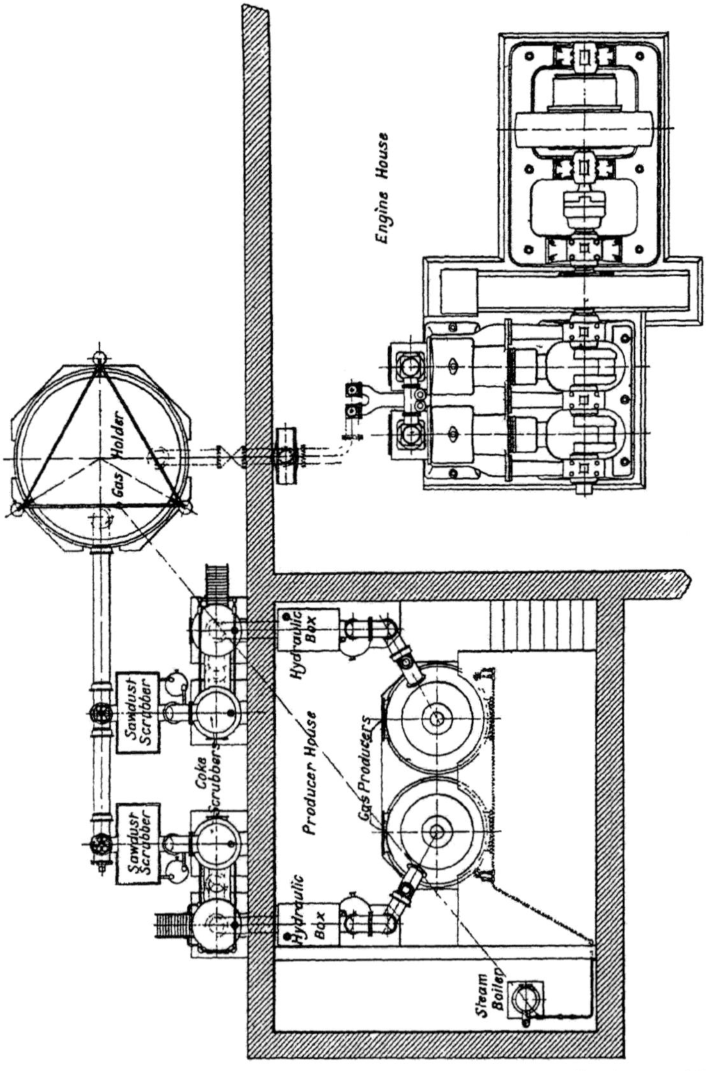

FIG. 82. Arrangement of 250 B.H.P. central station plant with two producers for continuous night and day working.

applied for combined heating and power installations with satisfactory results.

Water Gas.

By a variation in the method of working pressure plants using coke fuel, it is possible to obtain a gas practically free from the usual diluents, nitrogen and carbon dioxide, and composed almost entirely of carbon monoxide and hydrogen in equal proportions. Producer gas so made is technically described as "Water Gas."

The system of working a water gas plant is to blow alternately with air and steam; the air blow precedes the steam blow and the two are not mixed as in the plants hitherto described. The gas made during the steam blow only is collected and as the decomposition of the steam absorbs a large quantity of heat while the gas production goes on, it becomes necessary to renew the heat of the fire. This is done by interrupting the gas production and heating the fuel by burning a portion of it with air. The process is therefore intermittent and consists in alternate periods of gas production (introduction of steam) and heating of the fire (blowing with air). The former periods usually last 7 to 9 minutes, and the latter $\frac{3}{4}$ to 1 minute. In recent water gas plants, on the well-known Dellwik-Fleischer system, the efficiency of which as a water gas plant has caused it to supersede all previous methods, the air blow is arranged with a strong blast so as to ensure combustion to carbon dioxide and consequent generation of maximum heat for the steam blow, and in an intermittent process of the kind under consideration this is obviously the best system for efficiency. The gas made during the air blow (principally CO_2) passes up the chimney to waste; the water gas evolved during the steam blow is passed through a coke scrubber and from thence into a holder ready for use. The nett calorific value of the gas so made is about 280 B.Th.U.'s per cubic foot and the average composition is as follows:

	per cent.
H	51·0
CO	42·0
CH_4	0·5
CO_2	4·0
N	2·5
	100·0

Theory. The decomposition of steam always absorbs the same quantity of heat and it is therefore evident that any economy in fuel in a water gas plant must be effected by the method of heating the fuel. Until a few years ago it was assumed that in a fuel bed of sufficient depth for gas production, the only possible result of the blowing with air would be producer gas, *i.e.* a gas containing principally carbon monoxide, besides the nitrogen of the air. By such methods only about one-third of the heat value of the fuel was transferred to the water gas, while the rest was absorbed by the producer gas and unavoidable losses.

By the Dellwik-Fleischer process, however, during the blow a very large volume of air is introduced in a short time, and it has been found that in this manner a combustion principally to carbon dioxide is obtained. There is thus about three times as much heat available for the heating of the fuel bed as by blowing in the previously described manner, and it is possible to obtain a much higher yield of gas from the fuel and transfer as much as 70 to 75 °/₀ of the heat value of the fuel to the water gas.

If the whole of the heat of the combustion arising from the air blow was available for gas making, the system would be particularly efficient, in fact, in the latter case, efficiency would be 100 °/₀: naturally, however, a considerable portion of heat must pass away up the chimney as the fire is gradually brought to incandescence and this heat is lost. The object of the system is to obtain as much gas as possible in accordance with the reaction $C + H_2O = CO + H_2$, but as the temperature of the fire falls towards the end of the steam blow, this reaction naturally gives place to the alternative reaction $C + 2H_2O = CO_2 + 2H_2$. Moreover, as explained in Chapter 2, the reversible reaction tending to the destruction of the carbon monoxide into dioxide takes place as the temperature falls, and this explains the presence of the dioxide in the actual composition of the gas. The secret of the success of the Dellwik-Fleischer system lies in the careful manner in which the steam blast is regulated so as to secure uniform decomposition of the fuel without excessive sudden chilling.

If all the heat of combustion during the air blow were used to enable the former of the two reactions named above to take place the volume of combustible gas produced per pound of carbon would be 46·15 c. ft. which is naturally much less than in the case of ordinary producer gas. If the gas made were composed equally of carbon monoxide and hydrogen the calorific value would be 317 B.Th.U.'s per cubic foot. Of course, these theoretical results cannot be obtained, and it is a high tribute to the efficiency and practicability of the system we are now describing, that these theoretical figures can be approached so closely as to give a gas of the actual calorific value of 280 B.Th.U.'s per cubic foot on a yield of 31·5 cubic feet per pound of carbon.

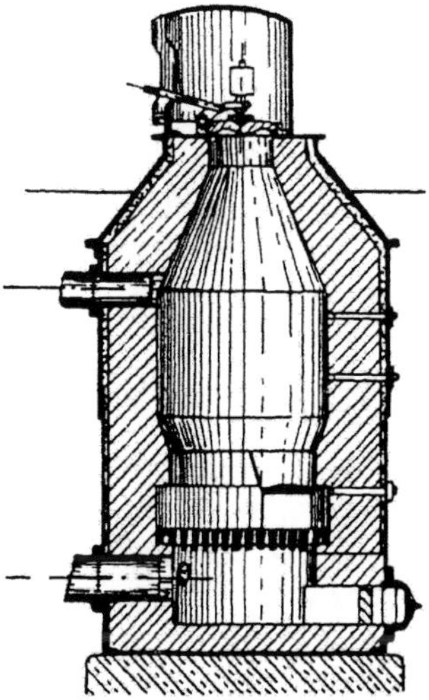

FIG. 83. Section through Dellwik-Fleischer water gas generator.

Description of the Dellwik-Fleischer plant. Fig. 83 shews the appearance of a generator. The surrounding sheet-iron shell is lined with firebrick. On a level with the clinkering doors is a grate supporting the fuel; below are ash doors for removal of the ashes. The air enters through the blast-valve, and the blow-gas leaves the generator through the central stack-valve, through which the fuel is also charged by means of a small coke wagon. There is one water gas outlet at the top of the generator and one below the grate, both connected with a three-way valve, through which the gas passes on its way to the scrubber. The gas-pipe is sealed with water in the bottom of the scrubber, where the gas is cooled and all dust washed out by the water running through the coke, with which the scrubber is filled. From the scrubber the gas passes on to a small holder, which equalises the flow of gas to the place of consumption. There is a steam pipe leading into the bottom and another to the top of the generator, as shewn in the general arrangement drawing Fig. 84.

The method of working is as follows: A fire having been built on the grate and the generator filled to the proper level with coke, the blast-valve is opened, and the fuel raised to a high degree of incandescence in a few minutes. Then one of the gas outlets—the upper one, for instance—is opened, the blast and stack valves being simultaneously closed by means of the gearing on the working stage. Steam is then admitted to the bottom of the generator and, passing through the bed of incandescent coke, is decomposed, forming water gas. A set of water gauges and a test-flame indicate the condition of the apparatus and the quality of the gas. When the temperature of the fuel has sunk below the suitable point, so that carbon dioxide begins to form in a larger proportion, the steam is shut off and the stack-valve opened, the gas valve being simultaneously closed. The blast valve is then opened for another blow of one to one and a half minutes. The periods of gas making are seven to nine minutes. For the next period of gas making the lower gas outlet is opened and steam admitted above the fuel. By thus reversing the direction of the gas making, the temperature of the fuel is equalised, causing less wear on the bricklining at any one point. The greater part of the coke being consumed by the action of the steam, the incombustible portions of the coke to a large extent disintegrate and fall through the grate as ash, while the clinkers on the grate are brittle and easily removable.

Properties of water gas. Water gas has a very high flame temperature, and is, therefore, very suitable for a number of manufacturing purposes for which producer gas cannot be used, and illuminating gas is too costly.

The principal manufacture based on the use of water gas is

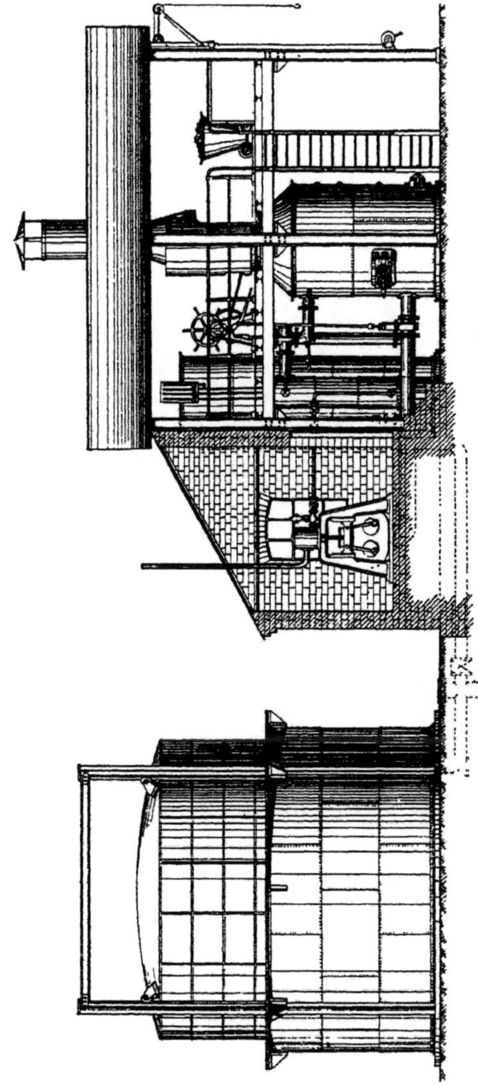

FIG. 84. General arrangement of Dellwik-Fleischer water gas plant.

plate welding, and a very considerable industry in this line has grown up, especially in Germany.

Water gas is further used for brazing and soldering, for heating furnaces of various kinds, annealing, tempering, etc., as well as for metal melting, either in crucible furnaces or open hearth furnaces. For open hearth furnaces it has the advantage of enabling the furnace to be quickly heated to a high temperature, and intermittent working is much easier than with producer gas.

Water gas does not offer any particular advantage over producer gas for power purposes, and as the water gas plant is more costly, we do not recommend it for this purpose alone, unless the gas is also used for other purposes. It is in the latter connection where combined heating and power are required that water gas is frequently applied for engine work.

SECTION III.

PRESSURE PRODUCERS FOR BITUMINOUS COAL.

CHAPTER 13.

DESCRIPTION AND GENERAL THEORY.

THE first gas producers such as those made by Siemens and others, were designed to use ordinary bituminous fuel, but as the gas evolved was, in such cases, used only for heating and furnace work, the presence of the tarry vapours in it which distil from such fuel was an advantage, for when burnt they added to the heating value of the gas. For power work, however, it is a *sine qua non* that the gas supplied to the engines should be *cool* and *clean*. Of course we use the term clean in a comparative sense; there is always a small percentage of impurities of one kind or another carried in suspension in producer gas. So far as the engines are concerned the impurities in the gas which cause the greatest amount of trouble are the tarry vapours referred to, which, condensing on the valve spindles, piston rings and igniter points, cause sticking of all these parts, and consequently forced stoppages of the engine at most inconvenient times. Generally speaking tar in the gas supplied to an engine leads to unreliability of working and most troublesome and costly cleaning of the working parts. As has been pointed out in Chapter 1, however, the percentage of volatile hydrocarbons which produce tarry vapours is high in all bituminous fuel and therefore for power purposes the cleaning out of these tarry vapours from the gas made from it necessitates costly scrubbing apparatus. Hence one of the chief distinctive differences between bituminous gas plants and those we have already considered for dealing with anthracite and coke, lies in these more elaborate scrubbers.

The producers also require important modifications, for the fuel to be treated has a large percentage of ash which generally tends to form excessive clinker in the producer. Hence there must be ample provision for poking the fire while the producer is at work so as to break up clinker which may arch across the fire, and further in most plants fire grates are altogether eliminated, the blast of steam and air being introduced through carefully designed tuyeres. The construction of various bituminous producers are considered presently and the features referred to above will be more fully understood by reference to the figures which follow hereafter.

It is usual to arrange the air supply to the producer from a Root's blower driven by a small steam engine which receives its steam from a small independent boiler. The exhaust steam from the engine together with any further steam obtained direct from the boiler as may be required, is allowed to mingle with the air blast on its way to the producer. The air and steam pass through the producer and react with the carbon of the fuel, and the gas made passes on through the coolers and scrubbers into a holder or sometimes direct to the supply.

Bituminous producers are seldom supplied of less capacity than 500 B.H.P., for against the cheaper fuel which they use must be set their increased first cost, increased amount of cleaning and attention and the disposal of ashes and effluents. In numerous cases where a cheap supply of anthracite or gas coke can be obtained, plants to use either of the latter fuels will be found to give cheaper gas up to 1000 H.P. than can be obtained with bituminous producers. Up to this power therefore each case must be looked into on its merits before deciding that a bituminous plant is necessarily the most economical. In isolated installations abroad, to which all the fuel has to be shipped and handled several times before it reaches its destination, it will seldom be found advantageous to use bituminous coal, as the cost of transport is so exceedingly high; in other words it is best in such cases to handle only the best coal, as with inferior fuel heavy freight charges are incurred in handling a large percentage of ash and volatile matter contained in such fuel which cannot be used at the engine.

The chief bituminous fuel producers before the British public at the present time are those by the Power Gas Corporation (Mond System), Messrs Crossley Bros., Openshaw, Messrs The Horsehay Co. Ltd. (Wilson System) and the Mason Gas Power Co. (Duff System) all of which have had a large measure of success.

The Mond Plant. Dr Ludwig Mond, F.R.S., was one of the first in the field with his well-known system which, designed primarily to produce gas for furnace work with the careful recovery of ammonium sulphate as a bye-product, has since become extensively adapted for power work both with and without bye-product recovery.

The usual Mond producer is shewn in section in Fig. 85 and consists of an inner shell carefully lined with fire-brick in which the fuel is reacted on by the air and steam blast. There is an outer jacket fitted all round the producer so as to leave an annular space of from two to three inches between the inner and outer shell and the blast is made to enter at the upper part of the jacket space and to pass through it on its way to the fire-grate at the bottom of the producer. A water lute is arranged below the fire-grate and as the jacket dips into this, a hydraulic joint is made which while leaving the bottom of the producer open to the lute so that ashes and clinker may fall freely therein, and be withdrawn therefrom, also prevents leakage of the blast. This arrangement of water bottom is now universally adopted by all makers for bituminous fuel producers.

The coal container at the top is made of considerable depth, so that the fuel in it becomes highly heated by the outgoing gases passing round the outside of the cast iron bell as they leave the producer on their way to the outlet pipe C. In this way considerable distillation of the coal takes place before the latter reaches the active zone lower down, as it is the custom to keep the container well filled with fuel while the producer is at work. The gases evolved by this distillation can only escape by passing down and through the upper part of the hot fuel in the producer, whence they mingle with the outgoing gases as they leave, and this is purposely arranged as the products of the distillation referred to are the light hydrocarbons which con-

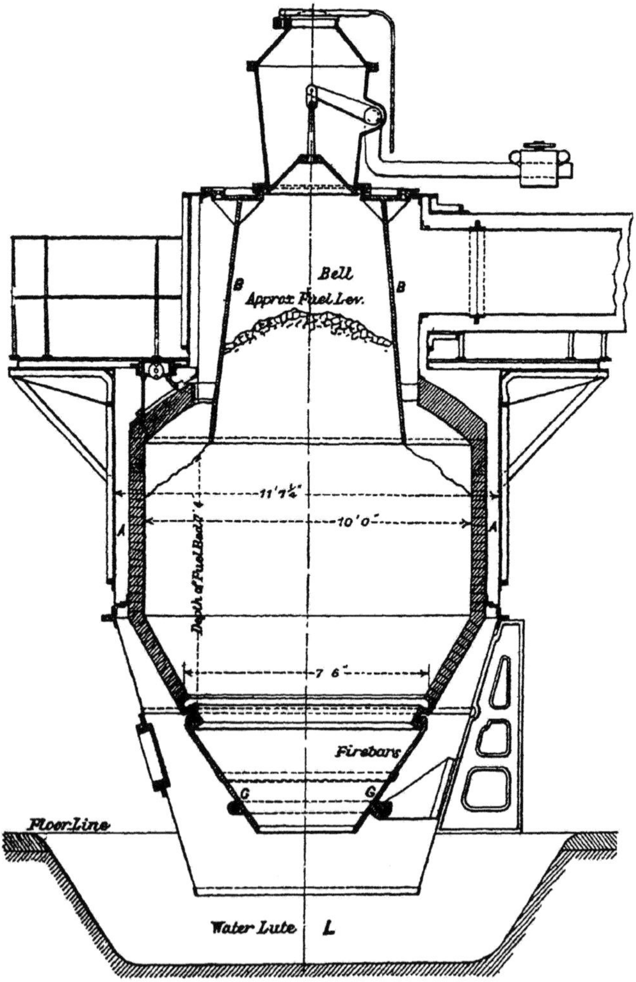

FIG. 85. Section through Mond producer.

dense as tar. It is thought that by these means a large proportion of these tarry vapours are decomposed into fixed gases under the influence of the high temperature of the hot fuel through which they are compelled to pass, but chemists are by no means agreed on this point, and in any case it is certain that most of the gases referred to are unaffected by any such temperature as those under consideration, unless there is sufficient free oxygen present to burn them.

It has been usual to work Mond producers with a large excess of steam (about 2½ pounds of steam for every pound of coal reacted on in the producer), this being considerably greater than is necessary for preventing the formation of clinker in the furnace. The low temperature of the latter which results from this, prevents the destruction of ammonia (NH_3) which is formed from the nitrogen in the fuel and which is afterwards used in producing ammonium sulphate.

For the general reasons given in Chapter 2, excessive steam in the producer is against efficiency, and as only 20 °/ₒ of the quantity of steam mentioned above is decomposed, it follows that there is a large amount of steam present in the gas leaving the producer and which if condensed out in the scrubbers and coolers would necessitate a much greater volume of cooling water and consequent excessive heat losses. The recovery of the waste heat in this surplus steam contained in the gas passing from the producer has therefore been most carefully studied and the hot gases are firstly passed through a so-called superheater consisting of a number of vertical pipes in series which have an inner and outer shell with annular space between. The hot gases pass through the inner tubes in one direction and the air and steam blast is driven through the annular space over these hot inner surfaces in the opposite direction. In this way part of the sensible heat of the hot gas and latent heat of unused steam mixed with it is transferred to the blast, and the temperature of the gas as it leaves the superheater is reduced to about 200° C.

After leaving the superheater the gas passes through a rotary washer in which the water used catches up a further portion of heat, reducing the temperature to from 90° to 100° C. In addition a great deal of tar and dust are here removed from the gas. Thereafter, when ammonia recovery is not attempted, the gas is further cooled in long vertical cooling towers wherein it meets with a downward stream of water as it passes up, and after the cooling towers a final cleansing takes place in sawdust scrubbers. In some Mond plants the cooling water used in the cooling towers, which becomes heated on its passage through the same, is returned to the top of air heating towers. This warm water is allowed to fall through the latter where it meets the ascending air blast and an exchange of heat takes place, the air in addition becoming partially saturated with vapour. This arrangement is shewn in Fig. 86.

On comparing this plant with the others which will presently be described, it will be seen that apart from details of construction in the various producers, the Mond system of heat recovery from the gases leaving the producer is much more elaborate than that usually adopted by other makers. In considering its comparative merits, it must be borne in mind that the general Mond design is the outcome of a plant arranged primarily for the recovery of the ammonium sulphate, the production of which necessitates a large excess of steam in the producer. As only a fraction of this steam can be decomposed, it follows that the greater portion of it appears unused in the gas evolved from the producer, and under these conditions more elaborate devices for recovering the large amount of heat represented by this unused steam may be efficiently introduced as compared with a system using less steam in the producer. On the other hand it has been found in practice that the recovery of ammonium sulphate cannot be profitably undertaken with a plant of less than 2500 H.P. and as by far the greater number of plants required in this country for factory driving are of smaller size, it follows that the consideration as to excess of steam and the after-recovery of the waste heat from the unused portion of it to which we have referred, do not apply.

The question of the proper amount of steam in relation to the air supplied to the producer, and the coal consumed in it from the point of view of securing the highest thermal efficiency as a gas producing system, therefore arises, and we have specially described the Mond system at this stage in order that our present reference to a striking series of experiments, carried out by Messrs Bone & Wheeler, on the effect of various proportions of steam in producer practice may be fully appreciated.

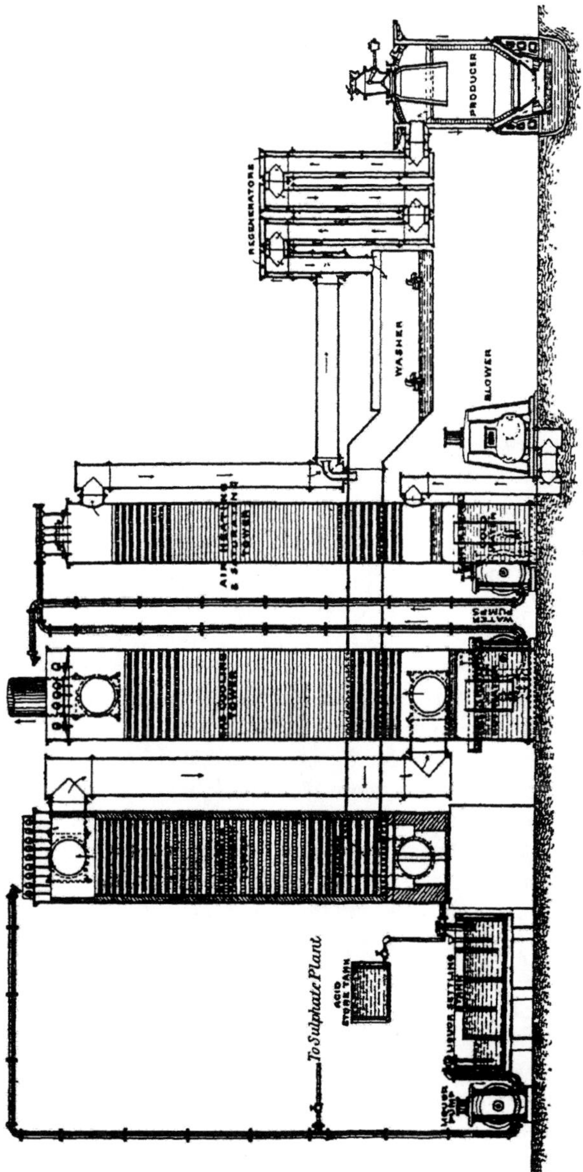

FIG. 86. Sectional arrangement (diagrammatic) of Mond plant with ammonium recovery.

Reactions in the producer. The general nature of the reactions which take place in a gas producer between the carbon contained in the fuel and the steam and air blast has already been stated in Chapter 2, and the same theoretical considerations may be held to apply to bituminous fuel producers. It was pointed out in Chapter 2 that the combination of reactions in Table 8 are most favourable to high efficiency since the gas produced under the conditions named is of the highest possible calorific value, but more particularly because a working temperature high enough to ensure these reactions taking place also prevents the destruction or decomposition of the combustible gases in the producer after their formation (see p. 29). Broadly speaking, we should argue from this that a bituminous fuel producer ought to be worked at a temperature just low enough to prevent the formation of excessive clinker and that no more steam should be used than is necessary for this purpose.

Messrs Bone & Wheeler's experiments* on the effects of varying quantities of steam in the air blast. These experiments were carried out in 1906 on a Mond plant of about 2500 H.P. in the aggregate, and consisting of two producers with heat regenerative and cooling apparatus similar to that previously described in this chapter. Their object was "to determine the influence of variation in the proportions of air and steam in the blast upon the composition of the gas, its suitability for furnace operations and upon the general and thermal efficiencies of the producers." The investigations comprised a series of five trials, each extending over a full working week, with the plant supplying washed gas both for gas engines and for re-heating and puddling furnaces, all working under normal conditions. The only difference between the conditions from week to week was in the relative proportions of steam and air used in the blast for the producers, which were varied by raising the steam saturation temperature of the blast in five equal stages from 60° to 80° C. In other words the amount of steam supplied with the air corresponded to saturation of the air at temperatures of 60°, 65°, 70°, 75°, and 80° C. respectively. These temperatures correspond to 140°, 149°, 158°, 167° and

* *Paper read before the Iron and Steel Institute, May* 1907.

176° F. and the weight of water per pound of air at saturation may be obtained from Fig. 13, p. 42.

It is further stated in connection with these experiments that:—"Suitable arrangements were made for

(1) the weighing of the whole of the coal charged into the producer, as well as that burned under the boiler, throughout each trial;

(2) the continuous sampling of the gas over a period of eight hours during each day shift, each sample therefore representing the mean composition of the gas generated in one day;

(3) the daily estimation of ammonia, tar vapour and sulphur compounds in the gas;

(4) the weekly collection and weighing of the tar deposited throughout the plant during each trial;

(5) the weighing of the dried ashes, clinker, and coke withdrawn from the water-troughs at the bottom of each producer;

(6) the estimation of the amount of soot and fine ash carried over by the gas and deposited in the mains during each trial."

It will thus be seen that the trials were most carefully and scientifically organised. "Collins Green washed nuts"—a very suitable non-caking coal—was used after being screened over a 1″ mesh and was found to contain from 3 to 6 °/₀ of moisture as supplied to the producer. Analyses of composite samples of this coal (dried at 105° C.) were taken during each week of the trials and were found to be as follows:

The mean calorific value of the dry coal was

Gross value = 3498 kilog. cent. units (= 13880 B.Th.U.) per lb.
Net value = 3362 " " (= 13340 ") "

Summarised Results.

The results in Table 21 are exactly what we should expect in view of the theoretical considerations laid down in Chapter 2. From Table 8 we find that the maximum weight of steam per pound of carbon which can be used if the highest efficiency is to be secured is 0·64 lb. The air per pound of carbon is

TABLE 20.

Week ending......	May 19	May 26	June 2	June 16	June 23	Mean
	per cent.	per cent.	per cent.	per cent.	per cent.	per cent.
Carbon	78·63	78·65	78·06	78·80	77·90	78·41
Hydrogen	5·18	5·37	5·22	6·50	5·30	5·51
Nitrogen*.........	1·37	1·39	1·37	1·40	1·40	1·39
Sulphur............	0·84	0·82	0·86	0·83	0·80	0·83
Oxygen (by difference).........	10·98	10·77	10·39	8·72	9·30	10·03
Ash	3·00	3·00	4·10	3·75	5·30	3·83
Volatile matter...	36·4	35·9	36·1	37·6	35·1	36·2

* This nitrogen, if wholly recovered as ammonia, would correspond to a yield of 147 lbs. of ammonium sulphate per ton of coal.

42·6 c. ft. or $\frac{42·6}{12·35} = 3·45$ lbs. (12·35 c. ft. of air = 1 lb.). The weight of steam per pound of air is therefore $\frac{0·64}{3·45} = 0·185$ lb. In Messrs Bone & Wheeler's experiments the best results were obtained when the steam saturation temperature of the blast was 60° C., where the air per pound of fuel was 36·95 c. ft., and the steam used in blast 0·455 lb. per pound of fuel, or steam per pound of air = 0·152 lb. 87·4 °/。 only of this was actually decomposed so that the net amount of steam used was

$$0·152 \times 0·874 = 0·132 \text{ lb.}$$

On referring to the analyses of the fuel, however, given in Table 20, it will be found that the proportion of carbon was about 78 °/。. From the results obtained from Table 8, Chapter 2, referred to above, we should therefore expect that the maximum amount of steam which can be used with advantage would be :—

$$0·185 \times 0·78 = 0·144 \text{ lb.}$$

It will be remembered, however, that these theoretical figures assume complete recovery of all the heat in the fuel, which of course is by no means possible in practice, so that we should say broadly on comparing the theoretical figure of 0·144 lb. with the corrected figure of 0·132 lb., corresponding to the actual amount of steam decomposed with a steam saturation

TABLE 21. *General Summary of Results and Conclusions.*

The principal results of the five trials may be summarised in one table, thus:

Steam saturation temperature (Centigrade)	60°	65°	70°	75°	80°
Coal consumption at producer, cwts. per hour	16·51	14·26	13·88	14·32	13·21
Coal consumption at boiler, cwts. per hour	2·0	1·8	2·38	2·87	3·35
Coal consumption for blast steam			0·38	0·87	1·35
Total carbon losses per cent.	5·8	7·8	8·1	7·1	8·4
Mean percentage composition of gas obtained: Carbon dioxide	5·25	6·95	9·15	11·65	13·25
Carbon monoxide	27·30	25·40	21·70	18·35	16·05
Hydrogen	16·60	18·30	19·65	21·80	22·65
Methane	3·35	3·40	3·40	3·35	3·50
Nitrogen	47·50	45·90	46·10	44·83	44·55
Total combustibles	47·25	47·10	44·75	43·50	42·20
Calorific value of gas (kilog. cent. units per cub. ft. at 0° C. and 760 mm.), gross	46·77	46·74	44·74	43·37	42·73
" " " " " " " " net	43·60	43·32	41·14	39·65	38·69
Yield of gas, cub. ft. per ton	138250	134400	141450	145800	147500
Pounds of steam in blast per pound of coal gasified	0·45	0·55	0·80	1·10	1·55
Percentage of steam decomposed	87·4	80·0	61·4	52·0	40·0
Cubic feet of air, at 0° C. and 760 mm., in blast per pound of coal gasified	36·95	34·9	36·8	36·9	37·1
Ratio of $\frac{\text{oxygen from steam}}{\text{oxygen from air}}$	0·50	0·62	0·65	0·75	0·80
Ammonia in gas as pounds of ammonium sulphate per ton of coal	39·0	44·7	51·4	65·25	71·8
Efficiency ratios in actual trials (1) including steam for blower-engine	0·778	0·750	0·727	0·701	0·665
(2) including steam for blower-engine and washers	0·715	0·687	0·660	0·640	0·604

temperature of the blast equal to 60° C., that no increased advantage is likely to accrue if the amount of steam is further cut down. Whether all producers can be worked with as little steam as that corresponding to air saturation at 60° C. must depend to a large extent on the character of the fuel and the proportions of the producer for the amount of fuel to be gasified. If the producers have to be forced a large excess of steam will be required to prevent the too rapid formation of clinker and under these conditions the best results cannot be obtained. Again, some fuels are much more likely to form clinker than others, and the amount of steam supplied to the blast must in many cases be almost entirely regulated by this consideration. It will be seen from the results of Messrs Bone & Wheeler's experiments, however, that a great gain in general efficiency follows on the cutting down of the steam from 1·55 lbs. to 0·45 lb. per pound of fuel gasified, and that the gas obtained when the latter amount is used is of a richer quality and containing a larger proportion of carbonic oxide as against hydrogen. This is of importance, both for heating and power work, because of the higher furnace efficiency of the monoxide for furnace work, whilst for engine work a gas rich in free hydrogen is always apt to cause "preignition."

The results in Table 21 also bear out the calculations given in Chapter 2 in respect of the yield of gas which increases as the amount of steam used increases, but this greater yield is not sufficient to counterbalance the loss of efficiency following on the destruction of carbon monoxide in accordance with the reversible reaction, $CO + H_2O \rightleftarrows CO_2 + H_2$, which tends from the left to the right hand side as the temperature falls through excessive steam being used.

If ammonia recovery be part of the system and the gas be required for heating and power it is suggested that the best all-round result will be obtained with the steam saturation temperature of the blast at from 65° to 70° C.

The preheating of the air blast by contact with hot surfaces, the temperature of which is maintained by the sensible heat of the outgoing gases leaving the producer, is calculated to increase the efficiency of the system as the heat transferred to the air blast in this way would otherwise be lost. The higher the temperature of the air blast, the greater is the quantity of steam which can be decomposed without loss of efficiency, but in practice it is doubtful whether the cost and complication of any elaborate preheating apparatus are justified by the increase of efficiency obtained. For this reason many successful makers dispense with preheating altogether.

Having discussed the general reactions in the producer and the effect of varying proportions of steam and air attention must now be turned to the question of the kind of fuels which can be handled in such producers.

Bituminous fuels for use in producer gas practice. It is not found easily practicable to use all bituminous fuels in producer practice, for though their chemical composition does not greatly vary, a distinct physical difference is met with when they are subjected to heat. *Caking coals* fuse together on heating and form into large masses which in gas producers restrict the passage of the blast and concentrate it into a series of distinct channels instead of allowing a uniform distribution over the whole area of the furnace. Under these conditions the necessary reactions cannot take place in a sufficient degree, and bad gas and general inefficiency result. *Non-caking coals* burn to ash without this fusion taking place and are obviously therefore more suitable for use in gas producers. In all districts more or less non-caking fuels are available and their use obviates a great deal of poking and general labour in working the plants. In the Midland Counties, Nottinghamshire Slack is generally used on this account, and in Lancashire, Collins Green coal is very suitable.

Gas plant makers usually base their guarantees on a non-caking bituminous slack of an average quality, containing not more than 15 °/ₒ of ash and moisture combined and not less than ·55 °/ₒ fixed carbon, and having a calorific value of from 11000 to 12000 B.Th.U. per pound.

A full analysis of non-caking Collins Green coal will be found in Table 20.

Peat as fuel. The question of peat as fuel has not yet arisen in a practical form so far as this country is concerned, as it cannot compete with bituminous slack when the latter is available at a reasonable price, for the simple reason that coal has a much higher heating value and naturally it does not pay to use the poor fuel when the better quality is to be had readily. An important factor to be borne in mind in connection with peat, however, is its richness in nitrogen, and ammonia recovery is therefore comparatively more advantageous. Peat producers in conjunction with recovery apparatus are on this account being increasingly employed abroad. From recent experiments by two German experts, Dr Frank and Dr N. Caro, on Italian peat used in a Mond plant, it was found that one ton of dry peat yielded 62850 cubic feet of gas of about 150 B.Th.U.'s calorific value and 118 lbs. of ammonium sulphate. The yield of the latter is exceedingly high, and these results from peat are exceptionally good. The calorific value of ordinary dry peat is about 7000 B.Th.U.'s per pound, or about one-half that of bituminous coal. The latter yields 130000 cubic feet, per ton, of gas of from 130 to 150 B.Th.U.'s with about 80 lbs. of sulphate; a ton of coal therefore gives fully twice as much gas but only about 0·675 the amount of sulphate which it is stated can be obtained from peat. It may be safely assumed that on the average it will cost as much to cut and dry a ton of peat on British bogs as to obtain a ton of bituminous slack delivered there, so that, for power purposes, coal is the cheaper fuel at the present time. Peat producers would require to be much larger for a given power and are therefore more costly in the first instance, but where a plentiful and cheap supply of good peat can be obtained there is no reason why, in conjunction with a recovery apparatus, they should not be advantageously employed. This may prove to be an important development in *the future application* of power gas producers.

CHAPTER 14.

CURRENT PRACTICE IN BITUMINOUS PRODUCERS.

In addition to the Mond producers of the type described generally in the previous chapter, large numbers of bituminous fuel producers of the various types we are about to consider have been successfully applied.

The Wilson Gas Plant.

The producer. This producer, designed by Mr Alfred Wilson, whose name has so long been identified with this class of work, is manufactured by Messrs The Horsehay Co., Ltd., Shropshire. A section through the producer is given in Fig. 88; the general arrangement of the plant is shewn in Fig. 87.

The producer is constructed with the ordinary water bottom to facilitate the easy removal of ashes and clinker while the plant is at work. It differs from the Mond and Crossley plants in having no grate whatever, the fuel in the producers being supported entirely on the piled up clinker and ashes at the bottom of the producer. The object of a grate in the producer is, as pointed out in Chapter 3, to distribute the air supplied as evenly as possible over the whole sectional area of the furnace. The designers of this plant consider, however, that it is a most difficult matter to keep the grates of bituminous fuel producers free from clinker when the plant is run continuously over several days, and if clinker adheres to a grate the successful working of the producer is completely or almost completely vitiated. In such cases, if the working of the producer must be continued, an excessive pressure of air blast must be employed to get the air supply through to the fuel at all, and on the other hand after the grate is passed, high air pressure tends to form channels through the fuel which are obviously bad from the gas making point of view, the air passing through without producing the required reactions in a necessary degree. These considerations constitute very strong arguments for entirely eliminating grates in bituminous plants, and in the Wilson producer successful working is further ensured by the addition of the special side poking holes in addition to those of the ball-valve type on the top of the producer, see Fig. 88. The side poking holes are kept gas-tight when not in use by a special arrangement of double closing plates, one of which slides vertically and the other horizontally.

The air blast is usually supplied from a Root's blower driven by a small steam engine, the exhaust of which is used to partially saturate the blast: any additional steam which may be required

Fig. 87. 600 b.h.p. Wilson bituminous plant.

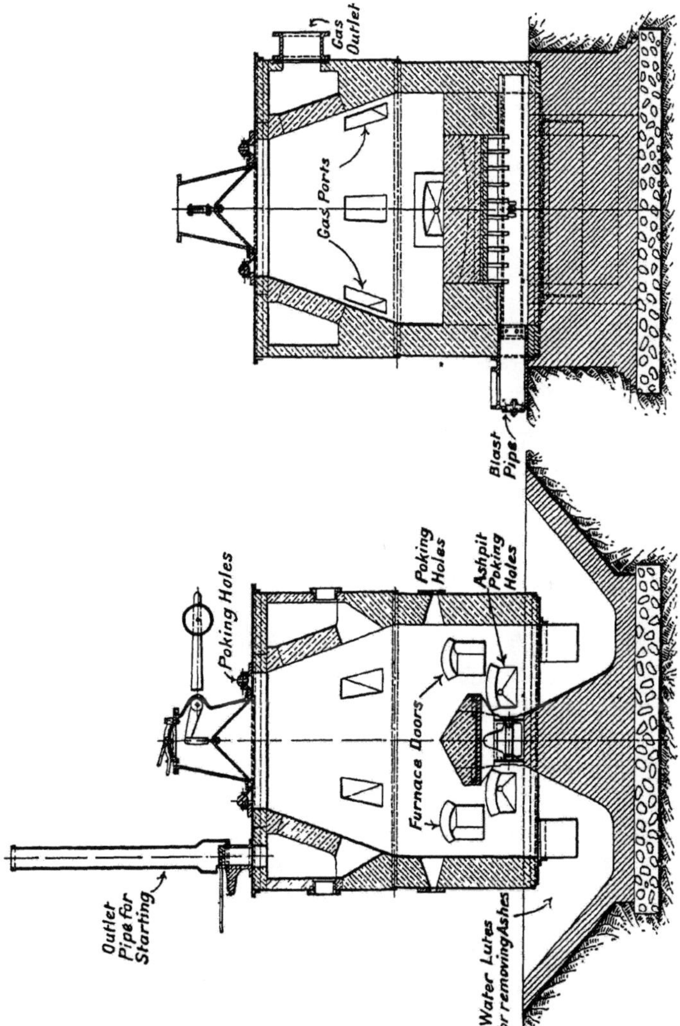

FIG. 88. Sectional elevations of the 'Wilson' producer.

is obtained direct from the steam boiler. The blast is brought into the inverted cone at the bottom of the producer and issues into the furnace from the tuyere openings at the top. A conical hood of firebrick is fixed over the tuyeres and serves the double purpose of assisting to distribute the blast and of preventing the fuel falling down and blocking the tuyeres. It is usual to bring the blast well up into the producer, so that there is the minimum tendency for the air to force its way through the bottom joints of the brickwork, and from thence up the space between the bricks and the outer iron shell of the producer.

No attempt is made to preheat the air on its way to the producer, as it is not considered that the gain in efficiency is sufficient to counterbalance the additional complication necessary to carry this into effect together with the increased first cost.

The coal is charged into the producers through the bell and hopper at the top about once every half-hour, and before the fresh fuel can pass into the active zone of the furnace it is heated in the container formed by the internal brickwork, which in turn is heated by the outgoing gases as they leave the producer. These gases pass from the furnace through a number of openings and from thence round the space formed behind the brickwork fuel container, after which they issue from the producer by one of the outlet branches. The object of the arrangement is to secure the distillation of the volatile hydrocarbons in the container and their conversion so far as possible into fixed gases by their passage through the hot fuel before they mingle with the other outgoing gases.

It will be seen that generally the producer is of an exceedingly simple design. Its conception is on sound lines and it works well in practice. The absence of any firebars or fire-grate, together with the scheme of top and side poking holes, is well calculated to secure continuous running with the minimum of trouble.

Cooling and cleaning apparatus. Mr Wilson has always been a strong advocate of the principle of atmospheric cooling and applies this system to his plants in most cases. The gas on

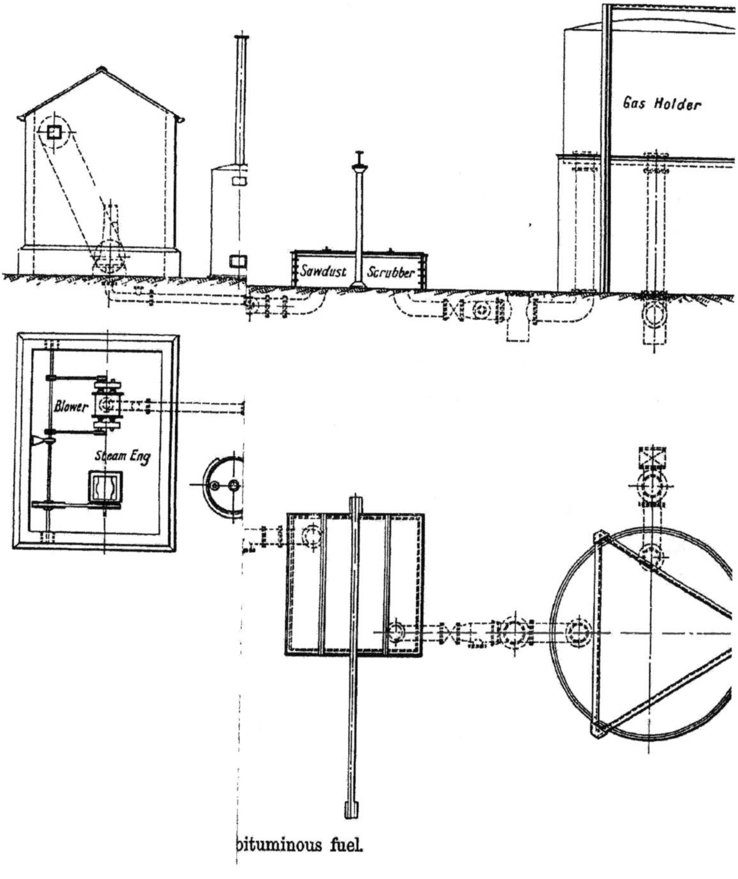

bituminous fuel.

leaving the producer is caused first of all to pass through a dust-catcher wherein the greater portion of coal dust blown through by the blast is arrested. This is effected on a similar principle to that employed in steam separators, viz. by introducing baffles and causing the gas to suddenly change the direction of its motion, with the result that the heavier particles carried in suspension, having a greater momentum, are precipitated. From the dust-catchers the gas passes direct. to the atmospheric coolers, the general arrangement being as shewn in Fig. 89.

These consist of long wrought iron tubes connected in series and of suitable number for the required duty. In order to increase the effective cooling surface an inner tube open to the air is introduced, and this is particularly effective seeing that a strong up-draught is naturally induced through it. As the gas passes up or down the annular space between the inner and outer tube of each cooler its heat is gradually dissipated to the atmosphere, and the object of the arrangement is to condense the gas to atmospheric temperature before it leaves the cooler *without using water*. It is considered that the introduction of water prior to this, saturating the gas with water vapour, converts each globule of this vapour into a vehicle for carrying tar forward and that under these conditions a final cleansing of the gas is rendered more difficult. In this connection it is a notable fact that atmospheric coolers on this principle are now being added to Mond and other plants.

There is no doubt also that the use of atmospheric coolers cuts down the water consumption of the plants enormously, and this is an important feature where a cheap supply of cooling water cannot be obtained. Another point is that as the tar is so largely condensed out of the gas without the use of water, it can be more easily collected and disposed of, and moreover the effluent water from the plant being less offensive causes less trouble in its disposal.

After being cooled in the manner just described, the gas is passed through a static washer of the "Livesay" type which is illustrated in Fig. 90. The gas after being broken up into fine streams by passing through the perforated tube—the perforation being immersed in water—bubbles through the latter and consequently loses a large portion of the remaining tar. The tar is

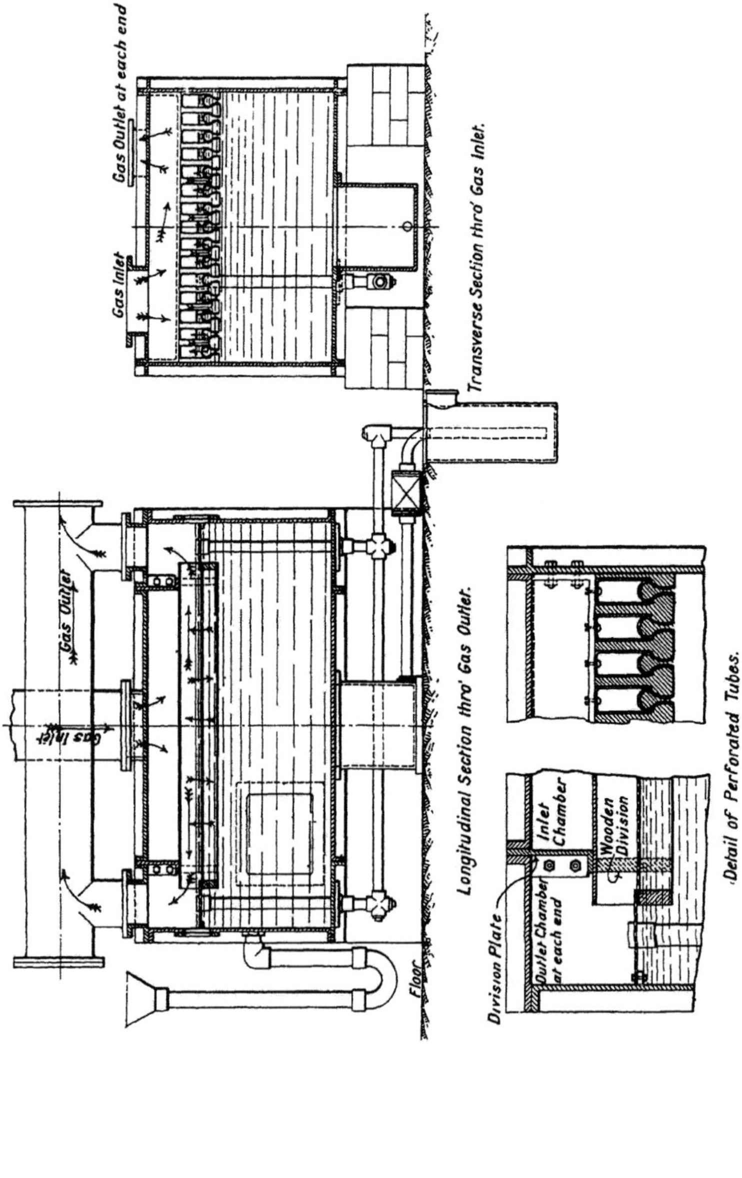

FIG. 90. Tar washer: Wilson plant.

drained off from the surface of the water in the washer from time to time. After leaving the washer the gas is finally dried and cleaned by being passed through a large sawdust scrubber and from thence passes to the holder ready for use.

An important point in connection with the plant proper should be noted, viz. that there are no working or moving parts to get clogged up. The only moving mechanism is confined to the blowing engine and blower, which are quite exterior to the actual gas producing apparatus. The significance of this feature will be appreciated, and we will have cause to refer to it again when we pass on to deal with the general considerations which should govern the design and arrangement of such plant.

The "Wilson" plant as made by the Horsehay Co. is a simple, workmanlike and efficient plant which has obtained a large measure of well-deserved success.

The Crossley Gas Plant.

Messrs Crossley Bros., Ltd., of Openshaw, Manchester, have manufactured bituminous plants during the last four or five years with a large measure of practical success. The usual design and arrangement of these plants are shewn in sectional elevation and plan in Figs. 91 and 92.

The producer. The producer is of a very simple design, consisting of a wrought iron cylinder and brick-lined shell with the usual water lutes at the bottom, to facilitate the withdrawal of the ashes and clinker whilst the plant is at work. A special feature of the producer is the rotary grate supported on a ball race which is shewn in section in Fig. 91 and described in principle on page 90. The blast connection is made into the cast iron cylinder upon which the rotating grate is supported, and the blast emerges from an inverted conical tuyere into the furnace. The coal is fed into the upper part of the producer in the usual way and the correct height of fuel in the furnace is maintained with the help of the cast iron coal container. No attempt is made to design the latter with the object of securing the passage of any distilled hydrocarbons through the hot fuel, and the consequent gain in simplicity is not the least important feature of this producer. Ball-socket poking holes are conveniently arranged on the top cover plate, and the means thus provided of poking vertically through the whole depth of the fire, together with the rotating grate referred

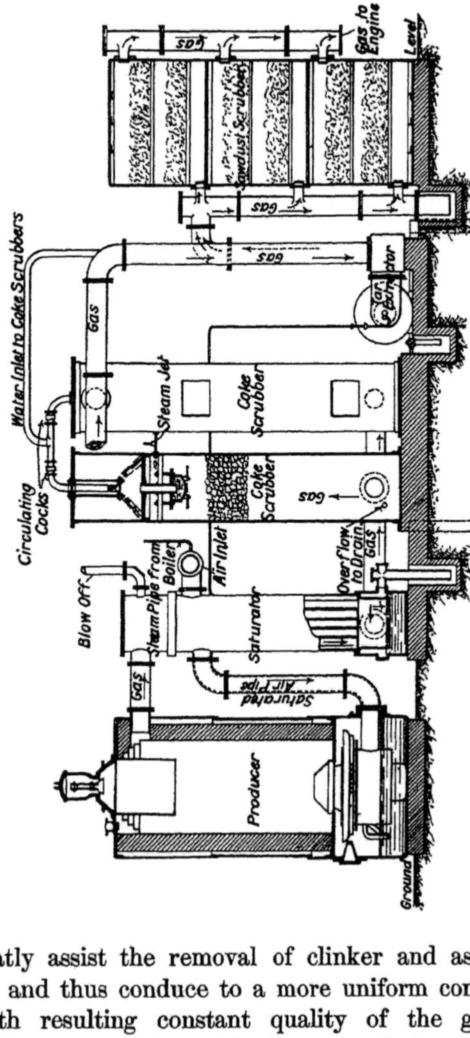

Elevation.

FIG. 91. Sectional elevation of Crossley bituminous fuel gas producer.

to, greatly assist the removal of clinker and ashes from the furnace and thus conduce to a more uniform condition of the fire with resulting constant quality of the gas produced. The careful manner in which these details have been worked out has overcome a great number of the difficulties experienced with grates in bituminous plants, to which we have already referred, and there is no doubt that this producer works well in the great majority of cases.

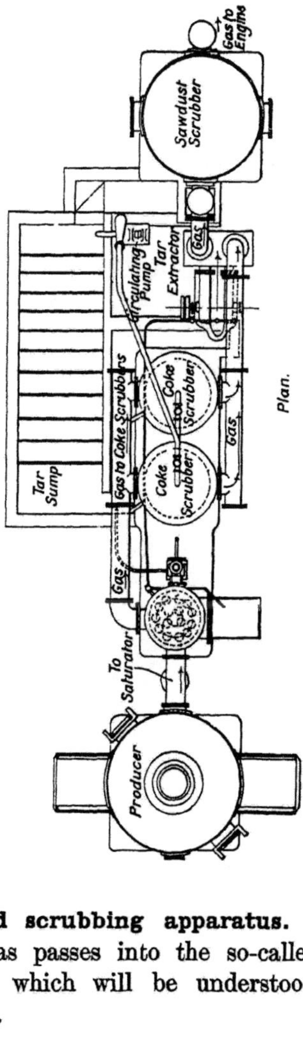

FIG. 92. Plan of Crossley bituminous fuel producer.

Cooling and scrubbing apparatus. After leaving the producer the gas passes into the so-called "saturator," the construction of which will be understood by reference to Figs. 91 and 92.

The gas passes through a number of internal tubes contained in a steel shell: between the top and bottom tube plates, into which the tubes are expanded, it is arranged that the air and steam blast passes on its way to the furnace and an exchange of heat consequently takes place between it and the hot outgoing gases passing through the tubes. As a result the air is preheated and water is vapourised to add to the steam supplied by the boiler, and further the gases are cooled very considerably before they come to the scrubber. The saturator also acts as a dust catcher, and consequently the bottom is arranged with a water lute which provides for cleaning at any time. The gases after passing through the tubes bubble through the water seal before passing to the outlet pipe, the object of the seal being to prevent the gas returning back to the generator at any time. After leaving the saturator, the gases pass through coke scrubbers, which catch a great deal of the tar and dirt in the gas, and before leaving these scrubbers it becomes cooled to about atmospheric temperature. By this time the remaining tar is well condensed and this makes circumstances favourable for its final separation in the mechanical washer. This is a patent centrifugal tar extractor, the special feature of which is to cause the gas to enter at the centre of the vane wheel on one side and to leave at the centre on the other side. The vane wheel is formed with a solid central disc (Fig. 93) which ensures that the gas takes the predetermined course through the fan from centre to periphery on the inlet side and from periphery to centre on the outlet side, and hence the work done in accelerating the velocity of the gas from zero to maximum is largely given back before it leaves the fan. Water is of course constantly introduced into this apparatus while it is at work, and the particles of water and tar are dashed against the stationary casing whence they drain into the sump. The final drying and cleansing of the gas is accomplished in the ample sawdust scrubbers before passing to the engine.

Other apparatus required to work the plant is as follows:

(*a*) A small steam boiler for supplying steam to the producer (sufficient steam cannot be raised in the saturator alone).

(*b*) A fan or blower for supplying the air blast at pressure.

(*c*) A small circulating pump for circulating the cooling water through the scrubbers.

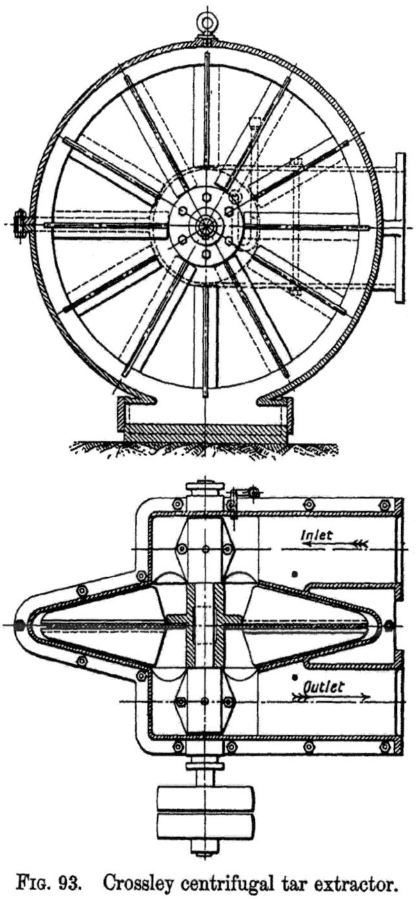

Fig. 93. Crossley centrifugal tar extractor.

(*d*) A main tar sump, see Fig. 92, in which the cooling and cleaning water is collected from the scrubbers and the tar separated by the skimming boards provided at intervals throughout its length. The water so cleaned is drawn away by the circulating pump and used over and over. A certain make-up is, of course, required, but this is stated not to exceed one half gallon per H.P. per hour.

(*e*) A small motor—either steam, gas, or electric—to operate the fan, tar extractor and pump. For this purpose a steam engine is usually applied as a matter of convenience.

It will thus be seen that there are distinct differences both in the producer and in the design and arrangement of the cooling and cleaning apparatus, between this and the "Wilson" plant already described. The general fundamental principles involved in both systems will be dealt with in Chapter 15, but we may conclude our description of the Crossley plants by observing that in their construction they embody the usual good practice of this eminent firm and that they work successfully where properly and wisely applied.

The Mason Plant.

The bituminous plants manufactured by the Mason Gas Power Co. Ltd., Atlas Works, Levenshulme, Manchester, are supplied in sizes up to 3000 B.H.P. for heating and power work, this firm being exclusively employed in their large new works with this class of work.

Their latest design of producer is shewn in Fig. 94. The generator is a simple, cylindrical, brick-lined wrought iron shell carried on four brick pillars between the four water lutes. The blast is introduced through a central pipe, and is spread at its outlet by a renewable cap. A circular slotted cone-shaped grate is arranged above the latter to distribute the blast evenly through the fuel, this grate being built up in segments as shewn in the plan view.

The gas outlet is taken from the side of the producer directly under the crown, and a drop valve is fitted to the bottom of the gas uptake below the shut-off valve, to allow of dust being removed while the plant is at work. Ball poker holes and covers of an improved design are bolted to the top plate of the producer through which the whole area of the furnace can be got at with the pokers. The side inspection and poking holes also greatly facilitate the working of the producer.

The fuel is fed regularly into the top of the producer through the customary bell and hopper at the top, and no attempt is made to secure distillation of the hydrocarbons prior to the fuel entering the furnace.

Cooling and cleaning apparatus. The general arrangement of the Mason plant as erected for use is shewn in Fig. 95;

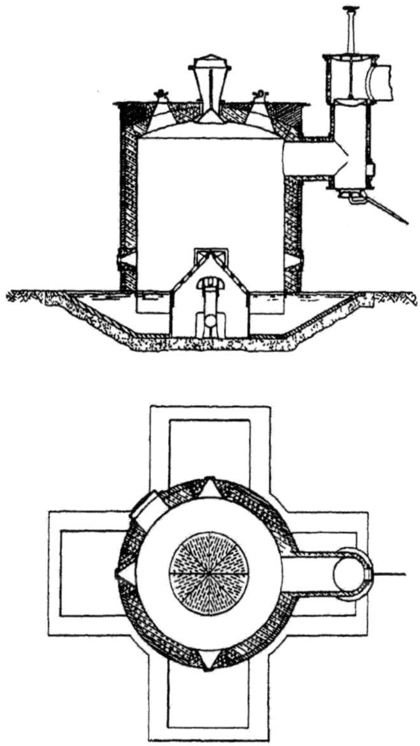

FIG. 94. Mason producer for bituminous fuel.

A is the Root's blower for supplying the blast to the producer, no attempt being made to preheat the air on its way thereto; B, B are a pair of producers in connection with which the coal elevator N, the conveyor O, and the bunkers P are arranged; R is the boot of the elevator into which the coal from the trucks is discharged.

The gas after leaving the producers passes first to the dust catchers C, these having a circular coned casing with the gas inlet arranged at one side to give a centrifugal action to the gas, and thus to separate the dust. A drop valve is fitted to the bottom of the catcher to allow of dust being removed while the plant is at work, and the outlet pipe to the annular atmospheric condensers is bolted to the top plate.

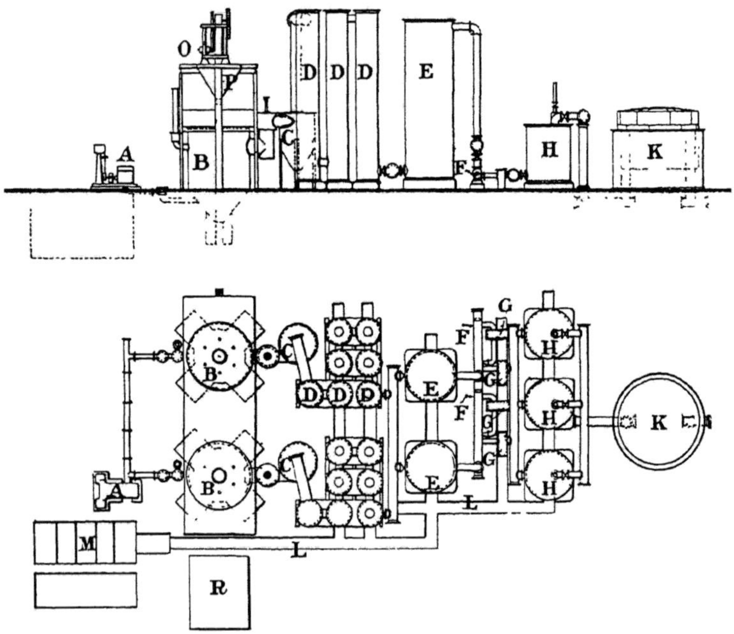

FIG. 95. General arrangement of Mason plant designed for continuous working.

A system of annular atmospheric coolers is arranged at D, these being similar to those described for the "Wilson" plant having steel casings carried on rolled steel joists with internal tubes, the gas passing between the inner and outer tubes, and being cooled before entering the coke scrubbers E. The latter are constructed with a circular riveted steel casing with gas inlet at the bottom and outlet at the top. Grids are provided for carrying the coke or other scrubbing material inside the casing, and water spray pipes are fitted beneath the top plate for distributing a fine spray of water over the coke.

After leaving the coke scrubbers the gas enters the mechanical washers F, which are of the centrifugal type fitted with a water spray near the inlet. After the gas has been passed through the first washer, it is sent through a tar extracting box G, fitted with an adjustable seal, and thence it is passed through a second centrifugal fan and tar extracting box similar to the first, thus providing very thorough means of eliminating the final traces of tar.

The final drying and cleansing is then carried out in the sawdust scrubbers H, which consist of circular steel casings fitted inside with several shelves on which sawdust, chips, wood, wool or peat moss litter are packed.

A small gas holder K in a steel tank is provided to act as an automatic gas regulator, a connection being arranged between the gas holder and blower for opening the escape valve on the blower whenever the gas holder is fully inflated.

A system of drain trenches L discharge into the tar settling tank M, containing the usual skimming boards for separating the tar from the purifying water. This settling tank is built of cement concrete with a pit at one side for periodically drawing off the tar, and after the tar has been removed from the purifying water the latter may be used again in the plant.

It will be seen from the foregoing description that the means provided for cooling and cleaning the gas are of a very complete character. The introduction of the atmospheric coolers enables the gas to be cooled without consumption of water for cooling purposes, and hence the only water used is for the purification of the gas. This ensures great economy in water consumption, which in the majority of cases is a vital consideration.

The purifying system of coke scrubbers, double centrifugal washers and sawdust scrubbers is much more ample than those usually provided in the other plants we have considered, and the gas discharged into the holder ready for use should be correspondingly clean.

Where the gas plant consists of more than one unit, omnibus pipes are arranged to connect up all the different parts of the purifying plant so that the gas from any one unit can be passed

Fig. 96. 2500 B.H.P. Mason plant.

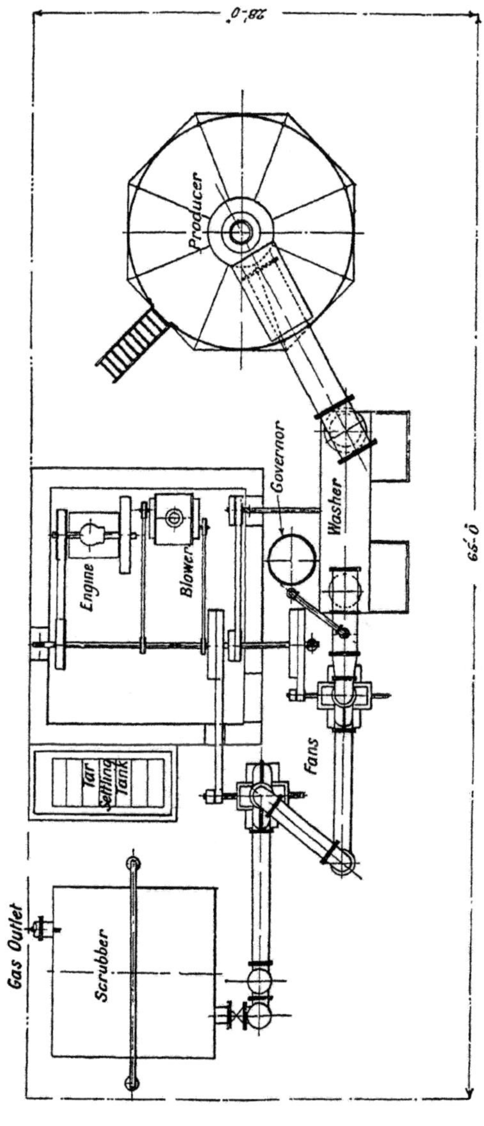

FIG. 97. General arrangement of 600 B.H.P. Mond plant.

through any portion of the purifying plant and not necessarily through those parts directly connected up to the gas generator by which the gas is being produced.

Generally the simplicity of the producer and the care with which the cooling and cleaning apparatus is designed and constructed are most important features in favour of this type of plant, which justify it being placed amongst the best bituminous coal gas plants at present offered for use.

Mond Plants for moderate powers. Fig. 97 represents a plan view of the general arrangement of a Mond plant of from 500—600 B.H.P. from which it will be seen that the gas on leaving the producer passes straight to the combined mechanical washer and cooler (see Fig. 98), in which the cooling water is intermixed with the gas by a revolving dasher driven from the steam engine line shaft. Leaving the cooler the gas is further passed through two washing fans placed in series and thence through an ample sawdust scrubber, after which it is ready for use in the gas engine supply mains. If the gas does not pass forward readily from the washer to the fans through the demand for gas supply being reduced, the pressure in the mains tends to increase as up to this point the blower is supplying the full quantity of air to the producer. Advantage is taken of this increased pressure to work a small gas-holder or governor which is connected to the gas main immediately after the washer: as this governor rises or falls it regulates the air supply to the producer through a suitable throttle valve, and thus the make of gas is controlled through a considerable limit. Otherwise the considerations governing the arrangement of the plant will be readily understood from the figure.

CHAPTER 15.

PRINCIPLES OF DESIGN.

GENERATORS.

Fixing of hydrocarbons in the producer. The practical utility of endeavouring to design the producer for fixing the hydrocarbons before they leave has long been a matter upon which expert opinion has been divided. In such cases the results obtained in practice under commercial conditions over a number of years usually settle the question, and considering the various issues on their merits from this point of view several unmistakable facts are presented. Practice has proved that generators in which the fixing of the hydrocarbon is seriously attempted require to be much more elaborate in design and construction than the simple producers considered in the previous chapter, and as a result their working is rendered more complicated and difficult, and their first cost is much greater. On the other hand it has not been possible to appreciably simplify the cleaning apparatus in connection with them, as the condensable tarry vapour still remaining in the gas leaving the producer is very considerable, and we understand that in the latter connection it has never been found practicable to reduce the amount of tar more than 30 °/o. Moreover scientific opinion strongly inclines to the view that the fixing of most of the hydrocarbons within the producer is not physically possible and that much of the so-called "fixing" is really a burning of the tar to CO_2.

Taking all these facts into consideration it is not surprising to find that most makers are gradually ceasing to design their producers to attempt any "fixing" of the hydrocarbons, with the result that they are a simpler job in every way.

We consider this course a wise one and must emphasise that the results hitherto obtained in the direction indicated do not justify the introduction of any complication in the construction of bituminous coal producers, which should accordingly be made on as simple lines as possible. This opinion, however, must only be regarded as applying to producers actually made and it must not be taken as an expression of finality in the matter. On the contrary, there are many competent minds at work on the subject, and doubtless before long their efforts will lead to practical success.

Grates. Another point upon which opinion is divided is in connection with the practical utility of fire-grates for supporting the fuel in the furnace. In considering this point, it must be remembered that one of the difficulties met with in the working of all such producers is the removal of the ashes and clinker while the producer is continuously in operation. The tendency is for these to arch across the lower part of the furnace and thus to seriously restrict the passage of the air blast, which is also prevented from being evenly distributed through the bed of fuel. Under such conditions air channels are formed through the fuel from bottom to top and the air rushes through these without properly reacting on the fuel, with the result that poor gas is formed. The presence of excessive free oxygen in the upper part of the producer arising from the air unconsumed also causes combustion to take place and consequently excessive heat is generated there with disastrous results.

The designer is therefore faced with the problem of the easy removal of ashes and clinker at all times and the proper distribution of the blast with the prevention of air channels through the fuel. A grate is *prima facie* desirable for assisting the distribution of the blast, but it should preferably be of the rotary type already described, and it ought to be kept well down from the hot zone so that it is practically protected by lumps of cool clinker etc., interposed between it and the active zone of combustion. If this be done there appears to us to be no real difference between the safety of allowing the ashes and clinker to rest direct on the ash pit than to be supported by a rotary grate, and there is no doubt that the latter affords a ready means of assisting in the stirring of the fuel and the consequent maintenance of an even fire.

Where the coal used is very fine or full of dust the advantages of a grate are even more marked, for it is extremely difficult in such cases to get any proper distribution of the blast otherwise, as air channels through fine fuel are very quickly formed.

We know of many producers fitted with grates which have been in successful operation for some time with results which warrant the foregoing views, and though it is difficult to generalise where circumstances differ so widely as in the case of gas plants, we consider that suitable grates should be introduced wherever practicable.

Area and depth of furnace. Here again practice differs widely, some makers cutting down the area of the furnace below the proportions laid down for suction producers, but in view of the large amount of ash and clinker in bituminous fuel we do not consider this wise. It must further be borne in mind that if a producer has to be forced, there is the tendency to generate great heat with corresponding excessive clinker so that the evils are cumulative.

For safe working we would advise a furnace area 30 °/₀ to 50 °/₀ larger for a given duty than is customary with pressure plants, for non-bituminous fuels (see p. 165), while the effective depth of fuel may with advantage be increased 20 °/₀.

Coolers.

It will have been noticed from the description of various plants given in the two previous chapters that there are great differences in the types of cooling apparatus respectively employed. There are, however, two broad classes; in one the cooling is done by transferring the heat to cooling water and in the other the gas is kept quite dry until its heat has been dissipated to the atmosphere.

Experience has shewn that both types possess respective advantages, and we are of the opinion that some discrimination should be shewn in deciding which is the best adapted for each particular case. In some of the gas plants we have considered, the gas after leaving the producer passes through a rectangular tank or washing box in which mechanically driven dashers or fans are continually throwing up fine streams of water through which the gas must pass, and as a consequence it is both washed and cooled. A general outside view of such a washer is shewn in Fig. 98.

Water lutes are provided at the bottom of the washer for removing dirt etc. at any time. So far as effecting an exchange of heat between the gas and water is concerned the comparative advantages of this arrangement are not apparent, as there is abundant experience to shew that such exchange of heat can be effectively carried out by passing the gas up a vertical cooling tower down which divided streams of water pass by gravity. From the point of view of cleaning the gas from tar, experience has also shewn that the mechanical operated tar extractors can best be introduced when the gas has been cooled and the tar vapours condensed.

FIG. 98. Mechanical washer for 3000 B.H.P. gas plant.

The shape of all the cooling appliances also require careful consideration in view of the possibility of an accidental explosion of back-fire in the producer at any time. These are not frequent and with stout cylindrical forms of coolers etc. no damage is done, but with the type of washer we have just described the long flat sides are easily distorted and can only be repaired at considerable cost.

Taking these points together, it would certainly seem that in most cases where the cooling is to be effected by water, ordinary vertical cylindrical scrubbers as in the Crossley plants, through which copious streams of water are continually descending, constitute a reasonable arrangement, and the only mechanical appliance required in connection with them is a small centrifugal circulating water pump.

Coming now to the atmospheric coolers, their chief claims are two: firstly, by cooling the gas without water they ensure a minimum consumption of the latter, and secondly, by condensing the tarry vapours without water, the effective cleansing of the gas afterwards is greatly assisted, as it is contended that mixed water vapour and tar vapour are more difficult to clean out of the gas than tar vapour alone (see p. 73). Regarding water consumption, this is not always the desideratum, as a large number of the works where gas plants are used are situated on the banks of rivers, streams or canals, and consequently, though important, the question of water consumption alone need not always determine the design. On the other hand there is no doubt that by the adoption of atmospheric coolers the *minimum* water consumption *is* secured and only about $\frac{1}{5}$ of a gallon per B.H.P. per hour need thereafter be used for washing purposes. For the alternative type of ordinary washer or cooler, where water is the chief cooling medium, it is argued that after skimming off the tar etc. and cooling the water, the latter can be circulated round and used again. Whilst by these means the water consumption can be reduced it must not be forgotten that the arrangement generally, with all this smelly water about, has obvious disadvantages.

When very fine coal is used in the producer, however, a large quantity of dust is carried into the coolers by the gas and this is more than the "dust-catchers" interposed between the producer and the coolers can usually precipitate. This dust is therefore carried on to the latter which it tends to clog up. Another difficulty which is experienced with fine coal is the liability, already referred to at the earlier part of this chapter (see p. 214), of channels being formed in the fuel through which the air passes to the top of the producer where it burns gas leaving the producer. This partial combustion may be carried on right into the atmospheric coolers, which consequently become excessively hot. The particular difficulty would not be so much felt if the gas was introduced into the presence of a copious supply of cooling water which could effectively prevent the propagation of flame.

Taking all the foregoing considerations together it would seem desirable in designing cooling apparatus to take account of the following factors:

(1) The general form of the coolers should be circular whenever possible so as to resist internal pressure such as may be set up occasionally by a back-fire in the gas plant.

(2) In the majority of cases the exchange of heat between the gas and the cooling water can be effected in an ordinary cooling tower without mechanical washers.

(3) Atmospheric coolers enable the minimum amount of cooling water to be used: whether their increased cost is thereby justified is a point to be decided on its merits in each particular case.

(4) With very fine fuel it is safest to rely on simple coolers fed with a plentiful supply of water.

When atmospheric coolers are adopted the required cooling surface (internal and external) is approximately 7 square feet per B.H.P.

Cleaning the Gas.

The removal of tar fog. We have already pointed out that for use in engines the gas must be practically free from tar as otherwise the valves, pistons etc. become stuck to an extent which prevents successful working. As a matter of fact the gas when used for engine work should not contain more than 0·1 gramme of tar per 100 cubic feet. In the types of producers we have already considered, centrifugal fans for dashing out the final traces of condensed tarry vapours have been applied in some cases and a stationary washer in others. It is important to know the relative efficiencies of these two methods and an impartial enquiry into the matter has recently been made by Messrs Clayton & Skirrow, who have fortunately published the results which they obtained.

Messrs Clayton & Skirrow's experiments*. These experiments were undertaken primarily with a view to arrive at more efficient means of removing tar fog from town's gas, as the presence of tar necessitates the more frequent renewal of the oxide in the purifiers. The first point in the investigation was to devise a method of accurately determining the percentage of tar fog in the gas before and after leaving the

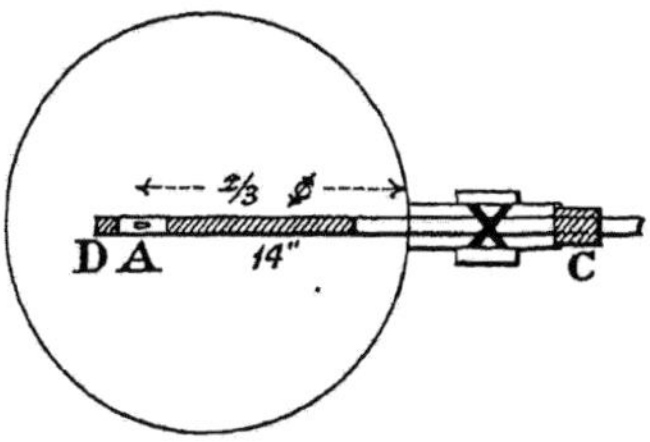

FIG. 99. Section through gas main shewing method of collecting gas sample.

washers, so that the relative efficiency of different types of the latter could be determined.

"The following is the detailed method of estimation of the tar fog in the gas. The main is tapped and a 1 inch pipe with a 1½ inch tap screwed in (see Fig. 99). This must be 1½ inches to allow the glass filtering tube to be put through into the main. The glass filtering gauge tube, ¾ inch outside diameter, is put through a rubber stopper C, and the length so arranged that the hole A is two-thirds (exactly) across the main. This is the point of mean velocity of the gas in the main. The filtering tube is closed at the end by a cork D, and the hole A is placed so as to face the current in the main, except in such cases where the current is upward and the tar so excessive that loss by drainage might be

* *Paper read before The Manchester and District Institution of Gas Engineers*, 1907.

feared—say, before the Pelouze and Audouin extractor. The diameter of the hole in A should be about ¼ inch. The current of gas should be taken through the tube at such a speed that the velocity of the gas through A is greater (say, twice as great) than the velocity of the gas in the main. The gas passes through 12 inches of very lightly packed cotton wool, which has previously been extracted with carbon bisulphide, care being taken that all the tube containing the cotton wool is in the main, and therefore at the same temperature as the gas. This prevents condensation. Generally, 20 or 30 cubic feet of the gas is taken. After the sample has been taken the tube is removed and any naphthalene which has condensed in the cooler and exterior part of the tube is carefully separated and removed, the cotton wool is pushed out into a soxhlet, and extracted with carbon bisulphide. After extraction, the bisulphide is evaporated off on the water bath, which is finally brought to 100° C. The flask is then quickly cooled, and a current of air passed through for half-a-minute. The cool flask is then weighed. Estimations on these lines were carried out at various works where we had the privilege of working; and the results are tabulated below."

Results obtained.

A. **Livesay washer** (Fig. 89). Note: The Livesay washer is of the type of tar-extractor already described in detail in connection with the Horsehay plant, see p. 200.

TABLE 22. Dec. 19th, 1906.

No.		Temperature	Tar per 100 cub. ft.	Purification, per cent.
1	Inlet	75° F.	11·26 gr.	
	Outlet	74 "	1·54 "	86·3
2	Inlet	73 "	12·92 "	
	Outlet	73 "	1·44 "	86·8
		Feb. 10th, 1907		
3	Inlet		10·5 "	
	Outlet		1·39 "	86·8

Tests were also carried out on the efficiency of the Livesay washers at another works, where the following results were obtained (Table 23).

TABLE 23.

No.		Temperature	Tar per 100 cub. ft.	Purification, per cent.
1	Inlet	86·5° F.	3·71 gr.	
	Outlet		0·44 „	88·1
2	Inlet	84·5 „	3·58 „	
	Outlet		0·54 „	84·9
3	Inlet	84·2 „	4·05 „	
	Outlet		0·52 „	87·2

From these results we see that the Livesay washer is not a very perfect form of tar-extractor; and although at works "B" the amount of tar left in the gas at the outlet from the Livesay washer is considerably smaller than is the case at works "A" yet we see that the percentage efficiency is practically the same in both cases.

B. **Pelouze and Audouin's Condenser** (Fig. 100).

This condenser, although not in extensive use in England, is very largely used in France and other continental countries, where the opinions expressed in its favour are very numerous, their general tenour being that by its use complete separation of the tar from the gas is effected, whilst the illuminating power, in the case of town's gas, is not impaired. They are also used in connection with blast furnace gas.

The passage of the gas through the apparatus is shewn by the arrows (Fig. 100); on ascending into the cylindrical chamber and passing through the perforations the gas is formed into jets, which strike against the solid surface sheet placed close to the perforated plates. In passing through the holes the liquid molecules are wiredrawn, and therefore brought into close contact with one another, the action being completed by contact with the solid surface upon which the tarry matter is deposited, from whence it flows down the surface of the plate into a tar underneath the apparatus.

Referring to Fig. 100, a general view of the apparatus is given in plan and elevation at A and B: the valves C are closed when the condenser is required to be put out of action for cleaning or for other purposes. A sectional elevation is shewn at D with arrows indicating the passage of the gas through the impact plates *e* and thence to the outlet: an enlarged view of these plates is given at E. F shews the balancing cylinder which automatically regulates the height of the cage of impact plates to suit the amount of gas passing through the apparatus:

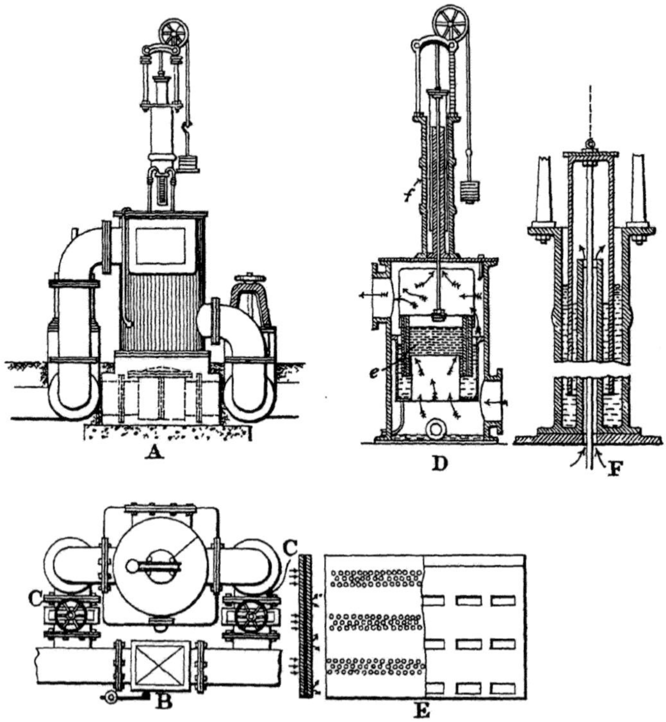

FIG. 100.

an increased quantity slightly raises the pressure and this lifts the balancing cylinder, which being connected to the cage raises the latter also out of the water seal and exposes more perforation through which the gas may pass.

The condensing cylinder, properly balanced, is capable of acting as its own regulator. For this purpose it moves in a hydraulic seal, which allows of the closing of those gas passages which are not required to be in action. When the pressure increases, *i.e.* when there is an increase of gas, the cylinder rises, as explained above, and a large number of openings are uncovered to allow the gas to pass through, thus adjusting itself to night and day or to winter and summer workings.

Messrs Clayton & Skirrow found this apparatus to be a most efficient tar-extractor as the results presently quoted shew, but it is quite unsuitable for producer gas as the cage becomes rapidly clogged. The principle involved, however, is both interesting and instructive.

Simultaneous tests were made at the inlet and outlet of the machines, the temperature and differential pressure being noted. This tar-extractor was one fitted with six plates—*i.e.*, the gas was subjected to three impacts during its passage through the machine. The size of the holes was $\frac{3}{64}$ inch, and the distance between the perforated plate and the slotted impact plate was $\frac{1}{16}$ inch. The capacity of the machine was 3 million cubic feet per 24 hours. The tests were carried out as described previously; and the following results were obtained:

TABLE 24. Nov. 14th to Dec. 17th, 1906.

Differential pressure	Temperature (Inlet)	Tar (Inlet), 100 cub. ft.	Tar (Outlet), 100 cub. ft.	Purification, per cent.
3·12 in.	73·3° F.	16·86 gr.	0·267 gr.	98·4
3·12 „	74·4 „	16·38 „	0·258 „	98·4
3·0 „	74·0 „	15·15 „	0·302 „	98·0
3·12 „	80–88 „	12·14 „	0·242 „	98·0
3·16 „	72·5 „	10·71 „	0·258 „	97·6

Tests were also made at the inlet to the purifiers after the gas had passed through the cyanide, ammonia, and naphthalene washers and the amount of tar found was as follows (Table 25):

TABLE 25.

No.	Temperature	Tar per 100 cub. ft.
1	60·6° F.	0·162 gr.
2	59·2 „	0·149 „

The above results shew that the tar can practically be completely eliminated by the proper use of a Pelouze and Audouin tar-extractor.

From enquiries made by the experimenters it was found that Pelouze and Audouin tar-extractors are often worked at considerably lower differential pressures than the above; and they determined to try the effect of different differential pressures, with the following results (Table 26):

TABLE 26.

Differential pressure	Temperature (Inlet)	Tar (Inlet), 100 cub. ft.	Tar (Outlet), 100 cub. ft.	Purification, per cent.
4·75 in.	72·5° F.	11·33 gr.	0·126 gr.	98·9
4·62 „	62·0 „	15·21 „	0·133 „	99·1
4·50 „	65·0 „	15·54 „	0·156 „	99·0
4·75 „	81·0 „	11·72 „	0·185 „	98·4
4·81 „	83·8 „	12·89 „	0·155 „	98 8
2·0 „	71·2 „	15·0 „	0·421 „	97·2
2·0 „	71·7 „	15·56 „	0·420 „	97·3
1·50 „	69·0 „	10·98 „	4·89 „	55·4
1·50 „	68·0 „	11·05 „	3·41 „	69·2

From the foregoing results it will be noticed that under a differential pressure of 2 inches (water gauge) an efficiency of over 97 °/ₒ was obtained.

C. **Crossley centrifugal tar-extractor** (see Fig. 93).

This apparatus was tested on producer gas and ran at 400 revolutions per minute, the diameter of the disc being 9 feet and the nominal capacity 5 million cubic feet per day, or about 200000 cubic feet per hour. The tests however were carried out up to 100000 cubic feet per hour only owing to the conditions prevailing at the time.

In experiments Nos. 2 and 3, the preliminary washers were bye-passed; and in experiment No. 4 one of the coolers was out of action, leaving more tar than usual to be dealt with by the fan. We see that the amount of tar at the inlet in experiments Nos. 2 to 4 is as great or greater than in the case of water gas.

TABLE 27.

Water used on fan. 30 gallons per hour.

No.	Speed of Gas (1000 cub. ft. per hour)	Temperature	Differential pressure	Revolutions per minute	Tar (Inlet), 100 cub. ft.	Tar (Outlet), 100 cub. ft.	Purification, per cent.
1	60				3·8	0·426	89·0
2	50	75° F.	0 ins.	412	13·8	0·458	96·7
3	50	84 ,,	0 ,,	412	26·5	0·690	97·4
4	100	83 ,,	0·6 in.	400	9·36	0·570	93·9

Conclusions. The high relative efficiency of the centrifugal type of tar-extractor disclosed by the foregoing experiments coupled with their proved efficiency and certainty in practice would suggest that their adoption is well justified. The power required to drive a suitable centrifugal tar-extractor is very small and the water consumption is low, while their continued positive action is not the least important feature in their favour. They should be well constructed and close attention should be given to the lubrication of the bearings and easy access for cleaning purposes: the speed of rotation should likewise be reduced as far as possible.

Design of Washing Fans. We have already referred on several occasions to the facility with which small globules of water act as tar carriers. The object of mechanically driven washing fans is firstly to closely mix a certain amount of water into the gas with the object of catching the tar, and secondly to separate out the tarry water thus formed from the gas as it passes through the apparatus. The first operation is effected by introducing a well broken-up spray into the inlet of the fan as close to the revolving centre as possible: the second is performed by the fan blades which project the tarry water against the stationary casing from which it is drained away to the tar sump or settling tank.

A standard motor driven 20 inch fan is shewn in Figs. 101 and 102 as made by Messrs W. J. Jenkins & Co. Ltd., Retford.

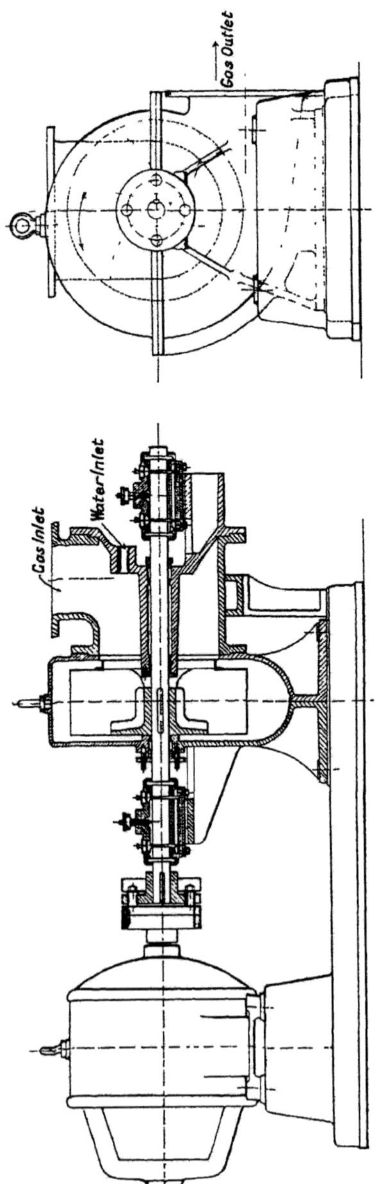

FIG. 101. 20 inch electrically driven washing fan, Jenkins' type.

The speed of such a fan is from 1700 to 2000 revolutions per minute at which it would wash 30,000 cubic feet of gas per hour against a pressure of from 4½ to 5 inches water gauge in the mains. Against a pressure of 3 inches water gauge this fan would pass 120,000 cubic feet per hour. A 36 inch double inlet fan of similar type would have a capacity of 480,000 cubic feet per hour at 4½ inches pressure.

The shape of the fan blades has a great influence on the washing effect, and on the pressure generated. For high washing efficiency the blades should curve forward at the tips,

FIG. 102. Jenkins' motor driven washing fan.

but this shape also produces high water gauge, which is not always desirable. When much water has to be used to clean the gas, more power is required with the blades curved forward than with a backward curve, and the proper curve to be adopted in any given case depends upon a compromise between these various factors.

For the thoroughly satisfactory cleaning of producer gas made from bituminous coal, it is often desirable either to arrange two fans in series or to fix two centres on one spindle and run them in a casing so arranged that the gas passes first through one end and then through the other. To pass the gas through one fan is sufficient for cleaning it of the worst of the tar, but for satisfactory cleaning when the gas has to be used in a gas engine two washings are necessary. Of course this doubles the final pressure, but it is very easy to take out the pressure by fixing plates sealed in water in a straining box, so that the gas has to force its way through the water and under the plates. This has the further advantage of additional cleaning of the gas.

FIG. 103. Belt driven washing fan.

It should also be borne in mind that a fan produces a difference of pressure between the inlet and outlet, and by throttling the inlet a vacuum can be produced in the fan so that the delivery can take place at the desired pressure either above or below the atmosphere within the limits of the fan, but this difference of pressure can of course be entirely obviated by arranging both inlet and outlet at the centre as in the Crossley fan.

The amount of water required to thoroughly clean producer gas in fans, depends very much upon the arrangements for cooling and scrubbing the gas when it leaves the producer before it enters the fans. A very common allowance is one gallon of water in each fan for every 80 cubic feet of gas passing or one gallon of water per 40 cubic feet of gas washed. This water can be used over again after the tar has been skimmed off, so that the actual water consumption is not a serious matter.

Sawdust scrubbers. Experience has shewn these to be necessary for the final cleaning and drying of the gas. In bituminous coal plants they should be large in capacity, and the chief points which regulate their design are ample sectional area through the sawdust trays so that the velocity of the gas while passing through the sawdust is as low as possible and the cleaning doors should be arranged in a convenient manner for the withdrawal and replacing of the sawdust. Their general construction will be understood on reference to the various illustrations already given.

General efficiency of Bituminous Coal Producers.

We have seen from the tests conducted by Messrs Bone & Wheeler referred to in Chapter 13 that the actual efficiency of a Mond plant may be as high as 77·8 °/$_0$ even taking the steam required for the blower into account. Under proper conditions similar efficiencies may be expected from the other producers described, and this fact confirms the view as to the doubtful expediency of introducing complications for the purpose of enriching the gas with the hydrocarbons which are otherwise condensed as tar, *i.e.* from a heat efficiency point of view. Constructionally the same considerations apply and lead us to the opinions we have expressed as to simplicity of design and arrangement.

CHAPTER 16.

MANAGEMENT, RESULTS, APPLICATIONS TO PRACTICE.

THE operations of starting and working bituminous coal plants are generally as follows:

Starting instructions. Having first made sure that the whole of the plant is in perfect working order, open the damper at the burning-off chimney on the top of the producer and close the top poking holes. Set the fire going as usual with sticks, oily waste, and coal, and, when it is burning well, open the air inlet valves to the producer and start the blower very slowly. Stir up frequently at the side and top poking holes and add fresh fuel until the furnace is charged to the top; look through the top poking holes occasionally to see that the air current has not worked holes through the fuel. If there is a blaze inside the producer, stir up either from the top or sides, until the hole in the fuel is filled and the blaze stopped. Now try from time to time to ignite the gas issuing from the top of the blow-off pipe, the speed of the blower being meanwhile increased until it is running at about half the normal full speed. As soon as there is a strong flame which will keep alight at the top of the blow-off pipe, the latter may be closed and the gas diverted through the coolers and scrubbers. Before this is done, however, the cooling water should be turned on, and it is usual to fit a blow-off cock immediately after the scrubbers and before the gas connection to the holder. This cock or valve is opened when the gas is first sent through from the producer and it is kept open for at least five minutes to allow any air or bad gas which may be leaking in the scrubbers to be driven off; the blow-off cock may then be shut and the connection to the holder or other point of supply may be opened. At this point the plant is ready for its regular work.

Working instructions. Put on the coal about every half hour, one or two hoppers full, never more than two hoppers full at one time, but sufficient to keep the fuel level right up to the coal container; in fact, just below the charging bell when in its lowest position. Whilst the plant is at work, stir frequently through the side poke holes, say every 15 minutes—THIS IS VERY IMPORTANT.—Put the bar in pointed down towards the hearth, and then depress the end held in the hand, so as to lift up the fuel and keep it free and open about the port holes on each side of the central tuyere. From the top poke holes keep the fire solid and break up any clinker that may form round the sides of the producer. It is very important that this should be attended to regularly, and that the whole mass of fuel in the producer should be kept gradually moving downwards.

The producer is kept clean by withdrawing a certain amount of ashes through the water lutes. This can be done whilst the producer is at work. With the shovel provided for the purpose fish out the ashes below the water equally on both sides of the producer, taking not more than a small barrow-load of ashes from each lute at one time, and also taking great care that the ashes are drawn from the whole of the grate area, and not at one place only. The operation should be done often enough, so that at the end of each fishing unburnt fuel begins to come. The producer should be cleaned at least once in the 12 hours, but experience of each individual case will shew how often this is necessary. At least once in the 12 hours all taps in drain pipes should be opened, and the pipes drained clear, and all luted drains should be examined to see that they are in proper working order. All water should be pumped out of the syphon pots and inlet and outlet valves to holder once every day.

It is important that the water in the bottom of the scrubbers or in the various seals should not be allowed to freeze.

The amount of steam turned into the producer along with the air blast must be adjusted so as to avoid excessive clinker. If more steam than this be used the gas made will be of poorer quality than it otherwise would be.

To stop the plant. First fill the holder, then

(1) Slow down the blower.
(2) Open the blow-off valve on top of producer.
(3) Close the gas outlet valve at producer.

(4) Stop the blower.
(5) Close air inlet valve at producer.
(6) Close inlet valve from holder.
(7) Close outlet valve from holder.
(8) Stop water supply to washers and scrubbers.
(9) Open the very small air inlet at the end of the tuyere.
(10) Open the two top poke holes.
(11) Close blow-off valve.

In this condition the plant can remain ready for starting up for 12 hours, or two or three days.

WORKING RESULTS.

As regards performance, the best series of authentic experiments would appear to be those carried out by Messrs Bone & Wheeler referred to in Chapter 13.

In addition the following information based on general results from several plants should be studied:

Coal consumption. The usual commercial guarantee is that from one ton of non-caking bituminous slack containing not more than 15 °/。 of ash and moisture combined, and not less than 55 °/。 of fixed carbon, and having a calorific value of 12000 B.Th.U.'s per lb., it is usual to expect a gas output of from 120000 to 130000 cubic feet with a calorific value 130 to 140 B.Th.U.'s per cubic foot.

Through the courtesy of Messrs The Mason Power Gas Power Co., Ltd., Manchester, we have been handed copies of a test carried out on a gas plant at Messrs Rowntree & Co.'s, York.

The following is a summary of the results:

Capacity of gas plant—560 B.H.P. in one unit.
The gas is used for driving two "Premier" gas engines.
Full load output—500 electrical H.P. (1622 amps. at 230 volts).
Output during 12 hours' test—4198 kilowatt hours,
= 5627 electrical H.P. hours,
= 6253 B.H.P. hours (estimated).

Coal burnt in producer—5510 lbs. of an average calorific value of 12410 B.Th.U. per lb.

Coal consumption—1·31 lbs. per kilowatt hour,
= 0·98 lb. per electrical H.P. hour
= 0·88 lb. per B.H.P.

Gas generated—134000 cub. ft. at 0° C. and 760 mm. per ton of coal, of an average calorific value of 150 B.Th.U. per cub. ft.

Efficiency of gas plant—Ratio of nett calorific value of the gas generated to the nett calorific value of the coal burned = 0·72.

In a usual way the consumption with ordinary slack in a bituminous plant should not be reckoned on as being less than 1·3 to 1·5 lbs. per B.H.P. hour, and this usually does not include the fuel required for steam raising purposes. In recent plants, however, it has been proved possible to burn the tar drained off from the extractor in connection with the plant in a special burner under the boiler instead of coal; this naturally effects a very considerable saving.

Water consumption. For steam raising purposes there should be available not less than 1¼ lbs. per B.H.P. hour, and when atmospheric coolers are adopted the only additional water required is one-fifth of a gallon per B.H.P. hour for cleaning the gas in the washer.

Cost of working and upkeep. Fuel and attendance are, of course, the two main items of the cost of working these plants, and it is very difficult at the present time to obtain accurate figures of what these really amount to. Makers contend that one attendant can easily look after a plant of 500 B.H.P. capacity provided that there is a little additional help at week-end times for cleaning purposes. For a plant of this size also there will be a charge equal to 6*s.* per week for replacing sawdust and coke in the scrubbers, while interest, depreciation and repairs must be reckoned at not less than 12½ °/₀. Generally speaking, users who have suction plants in operation and have also experience in bituminous plants are of the opinion that in this country there is very little advantage in applying the latter up to 500 or 600 B.H.P. at least, but, no doubt, improvements will be made in the future which will assist the successful application of bituminous producers, which are really only in their infancy.

Application to Practice.

The most remarkable installation of power gas in this country is the large public power supply of Mond gas in Staffordshire, about which so much has been written already that further reference here is unnecessary. Unfortunately, by the time this scheme was got into operation, the suction plant had come into use, and now constitutes a formidable competitor with all power gas distribution schemes. It is further open to serious doubt whether the principle of distributing ordinary producer gas over large areas is in itself a sound one, for it may be compared with the distribution of electricity at a low potential which is, of course, generally recognised as being wasteful from every point of view. To put the matter in another way, since producer gas contains between 50 to 60°/₀ of diluents, heavy capital and working charges have to be incurred for storing and distributing these diluents which are absolutely of no use to the consumer and for which he cannot afford to pay. This factor must of necessity cripple the revenue earning capacity of the comparatively enormous capital which has to be sunk to distribute a given number of heat units over a sufficient area to include a reasonable number of power gas users.

In the large chemical works of Messrs Brunner, Mond & Co. Ltd., Northwich, the Castner-Kellner Alkali Co. Ltd. and other modern chemical manufacturing establishments, Mond plants with ammonia recovery apparatus aggregating many thousands of horsepower have been in successful operation during the last few years, the gas produced being used both for heating and for power. The general economy which has attended their adoption has given the greatest satisfaction to those concerned. As indicated in the introduction to this book, there is also an increasing tendency to use producer gas for the heating and annealing of armour plating, etc. and many large installations are now in successful operation for this purpose. One of the chief advantages in such cases is the absence of oxidation of the plates during treatment, due to the fact that it is always possible to feed rather less air to the furnace than is actually required to fully consume the gas present; this ensures that the *whole* of the oxygen in the air is taken up by the gas and consequently none remains to oxidise the plates which are in the process of being manufactured. It is also obvious that as the gas and air supply are under such easy control, the exact degree of heat required can be regulated to a nicety.

Factory Driving. In ordinary factory driving where bituminous plants are used, the power required is in most cases too small to admit of the ammonia recovery being undertaken with advantage. Limits of space make it impossible to deal fully with the application of bituminous producers for all purposes, but two typical installations are shewn in Figs. 104 and 105 which will serve to indicate what is being done in this direction.

Fig. 104 shews a gas-driven electric-power station designed by Messrs Williams and Bridges of London for a large cement works in the Midlands. For elasticity of arrangement to meet varying load conditions at different times of the day, the aggregate of 600 B.H.P. is made up of four 150 B.H.P. National engines. It may be here mentioned that there is no loss of efficiency in adopting a number of comparatively small gas-driven units to make up a given total of power as the efficiency of gas engines does not rise with increase of power, as in steam engines; nor is there any increase of first cost, whilst on the other hand the greater elasticity and less risk of failure of the working parts in these small engines will be obvious. The gas plant is of the 'Wilson' type working with a very common variety of slack, but in spite of this the installation has run well and is giving every satisfaction. The consumption of coal at full load is less than one pound per B.H.P. hour. The general arrangement of the various parts of the producer plant and power house will be understood by reference to the figure.

Fig. 105 shews a 600 B.H.P. two-cycle gas engine recently constructed by Messrs Mather and Platt, Manchester, for driving a local cotton mill. It is working with gas made by a 'Wilson' type producer plant using Collins Green coal (see page 190) on a guaranteed consumption of one pound per I.H.P. hour. All the onerous conditions of absolutely steady and reliable running,

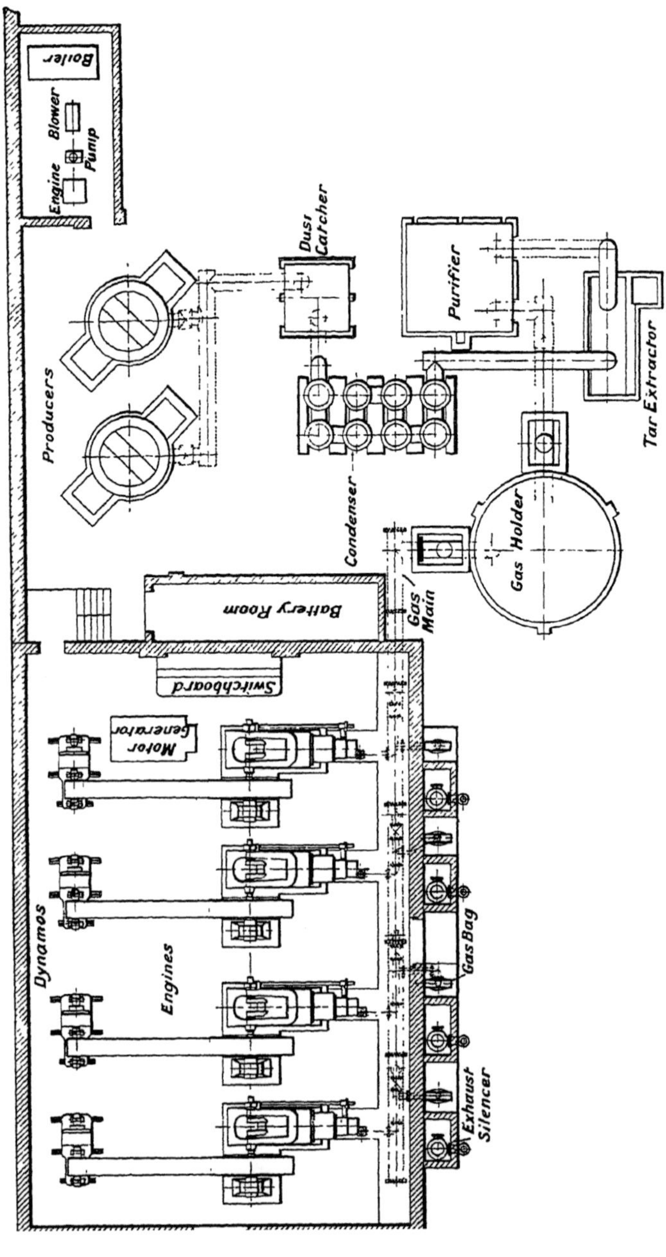

FIG. 104. 600 B.H.P. installation of National engines to 'Wilson' bituminous gas producer.

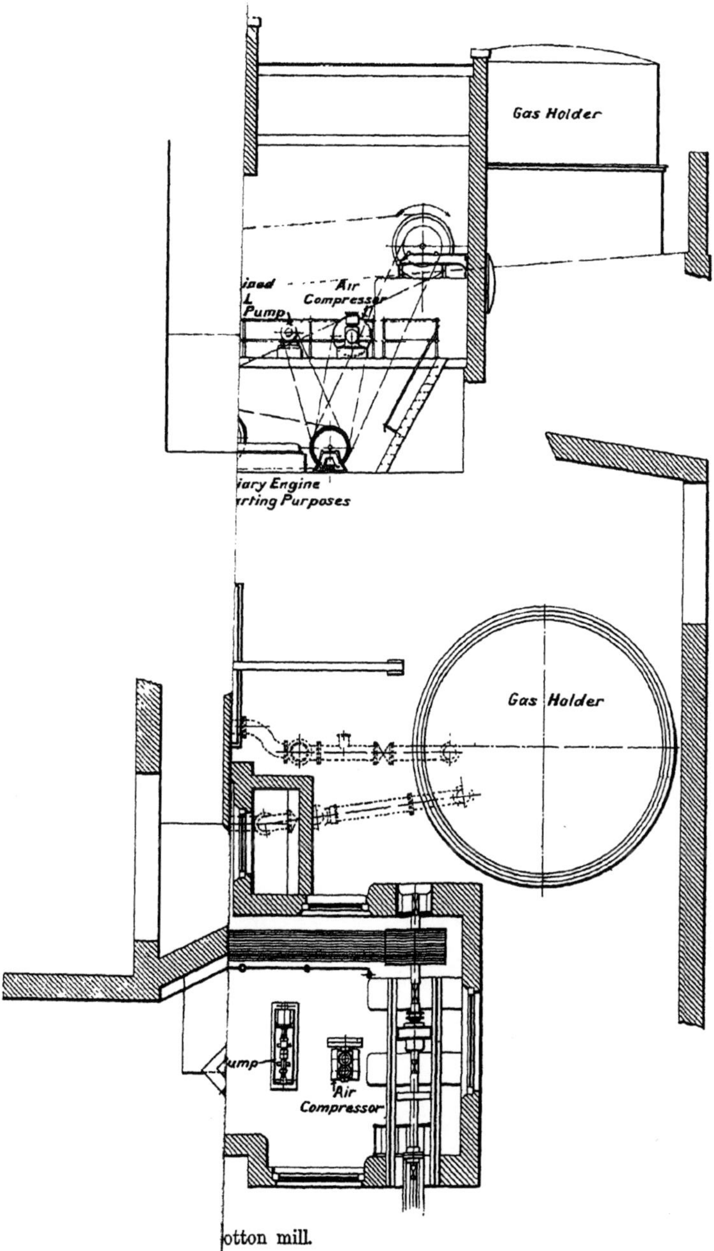

Gas Holder
ined
L
Pump
Air
Compressor
iary Engine
rting Purposes
Gas Holder
Pump
Air
Compressor

otton mill.

which are essential where textile machinery has to be driven, are being satisfactorily fulfilled, and we understand that the plant generally is giving every satisfaction.

Conclusions. It is realised by all who are intimately concerned with the design and construction of large gas engine and gas plant units that much yet remains to be done before the fullest economies, which appear to be reasonably possible, are brought within the region of actual practice. But the results which have actually been achieved within the last few years and the rapid progress which has been witnessed during that time are full of encouragement both to maker and to user. So far as large gas plants are concerned, greater simplicity and sounder constructional details are the chief points to which attention should be directed, though even meanwhile power users will find that the claims of gas power are so numerous and well founded that in the hands of a good firm they will have no difficulty in securing results which will well justify the adoption of this method of driving the power and heat they require.

AMMONIA RECOVERY.

When coal is heated a certain amount of ammonia (NH_3) is given off but if the temperature at which this heating takes place is such as is found in a gas producer under usual conditions, the ammonia gas would be destroyed by decomposition or combination with other gases. If, on the other hand, this destruction of the ammonia can be prevented and its recovery afterwards effected, it is a most valuable bye-product for which there is a regularly good demand. In practice this is done by keeping the temperature of the producer low through the use of an excessive amount of steam and the gas made is afterwards put in contact with a solution of sulphuric acid which readily combines with the ammonia in the gas to form ammonium sulphate.

If we refer to the analysis of the Collins Green coal given in Chapter 13, p. 190, for instance, it is there noted that the nitrogen in the fuel if wholly recoverable as ammonia would correspond to a yield of 147 pounds of ammonium sulphate per ton of coal. In recovery practice however it is not usual to expect more than from 70 to 90 pounds of sulphate per ton of coal consumed. Sulphate sells for about £12. 10*s.* 0*d.* per ton, so the value of an average yield say of 80 pounds is no less than 9*s.* per ton of coal burnt at the producer. This will probably represent the average cost of bituminous slack such as is used for gas-making at the present time, so that under suitable conditions the price obtained for the sulphate counterbalances the first cost of the fuel used, whilst in addition there will be a yield of about 140,000 cubic feet of producer gas per ton of fuel representing about 2,000 B.H.P. hours of energy. On the other side of the cost sheet, the increased first cost of the recovery plant and also the heavier working expenses have to be set against the factors referred to above, and there clearly must be a point below which it does not pay to trouble with recovery at all. Unfortunately up to the present time it has not been found worth while on plants below 4,000 B.H.P. gasifying less than 30 tons of fuel per day, so that it is only applied in the largest installations.

The chief recovery plants at work are those designed by Dr Ludwig Mond, F.R.S., and are generally arranged as in Fig. 86. In order to keep the working temperature of the producer sufficiently low to prevent the destruction of the ammonia as much as 2½ pounds of steam per pound of coal are passed through with the air blast. As not more than 20°/ₒ of this steam is actually decomposed, it becomes a matter for careful arrangement to recover the heat represented by the balance of this steam which is intermixed with the gas leaving the producer. Hence the two considerations which have influenced the design of the Mond recovery apparatus, as shewn in Fig. 86, are firstly to secure the ammonia, and secondly to take advantage of the latent heat of the undecomposed steam, in addition to the sensible heat of the gas, to preheat the air blast before it passes to the producer. The mixed gas and steam as they leave the producer are, therefore, firstly passed through the superheater described in Chapter 13, wherein the temperature is reduced to 90° C., and from thence into the mechanical washer where cleaning and further cooling take place.

Leaving the washer the gas is next passed through the vertical acid tower which is usually a wooden lead-lined erection filled with tiles arranged in such a fashion as to facilitate the breaking up of the gas into fine streams which intermingle with a descending stream of water containing about 4°/₀ of free sulphuric acid in addition to the sulphate in solution. The ammonia combines with the acid to form further sulphate which is carried in solution into a large well at the bottom of the tower and from which it is pumped to the top of the tower and caused to descend again through the ascending gas. This goes on until the sulphate liquor becomes sufficiently concentrated, after which it is pumped into evaporating pans heated by steam coils. In practice there is usually a constant feed of acid water kept running into the well at the bottom of the tower. The acid circulating pump for keeping the solution moving through the tower works at about the same rate as the feed, and hence it is possible to design the capacity of the bottom receiving well, so that if emptied quickly the time occupied in refilling by the acid feed will be sufficient to bring the liquor, which is meanwhile being continually passed through the tower by the circulating pump, up to the required standard for evaporation. The emptying pump which sucks from the liquor well and discharges into the evaporating pans is therefore comparatively larger than the others and works only when the well is full, and which it quickly empties as described. In this way the process can be carried on without difficulty by the ordinary class of attendant.

In the Mond system, the gas is further cooled after leaving the acid tower by being passed up a cooler through which streams of cooling water traverse downwards. An exchange of heat takes place between the gas and the water, and the latter consequently becomes somewhat heated. This heated water is further caused to pass down a third tower through which the air blast on its way to the producer is caused to pass, and an exchange of heat between the water and the air here takes place. The air blast then passes on to the superheater already referred to. Whether this elaborate system for the recovery of heat sufficiently affects the working efficiency of the plant is open to some doubt, and it would seem from some points of view that it might be eliminated altogether. On the other hand it is to be remembered that most of the Mond recovery plants at work at the present time are in operation day and night, week in and week out. Under these conditions even a fractional saving becomes of importance over a year's working.

Messrs Crossley Bros., Ltd. have recently introduced a modification of the foregoing system of recovery which they claim to be simpler and more efficacious. Their method is to use acid liquor in a mechanical washer of the Mond type through which the gas passes after leaving the superheater. There are a number of compartments in this washer, each with its own mechanically driven dasher, and the gas is caused to pass through these in series, the acid liquor being introduced so as to pass through neighbouring compartments in the opposite direction to that of the gas. An exchange of heat takes place between the liquor and the gas, the former becoming quite hot as it passes through the washer. It is consequently discharged to the top of a cooling tower arranged with the usual tiles, down which it trickles, and on its way it is met by an ascending air blast. The liquor gives up its heat to the latter which also becomes saturated to the limit of its temperature, and is afterwards passed through the superheater on its way to the producer. It is claimed that in addition to the simplicity of this arrangement, the mechanically driven dashers have the effect of so closely intermingling gas and acid liquor that less free acid is required and a greater yield of sulphate obtained. The free acid required is stated to be only one half of one per cent. and the yield as high as 100 pounds of sulphate per ton of fuel. To prove whether these better results are those which may be expected under ordinary working conditions further evidence will be necessary, especially as the ammonium liquor will be so much dirtier under this arrangement.

Bye-product recovery seems likely to play an important part in the future owing to the increasing attention which is being given to the utilisation of the less rich and partially formed fuels, such as peat, lignite, etc. These latter are comparatively richer in nitrogen than ordinary coal and this factor makes recovery correspondingly of greater importance.

APPENDIX A.

COKE OVEN GAS.

As power gas engineers have frequently to deal with the utilisation of the surplus gas from bye-product coke ovens, the following detail notes will be found convenient for reference:

The gas produced in these ovens has a composition similar to that of illuminating gas, as the coal undergoes a similar distillation to that carried out in gas works. Tar, ammonia and benzol are usually recovered in a special bye-product recovery plant. Sulphur compounds and carbon dioxide are not separated from the gas, which also contains a small quantity of creosote oil and other tarry matters and impurities carried over mechanically from the recovery plant. We might therefore describe the gas as an impure and inferior quality of illuminating gas.

The gas is drawn from the ovens by compressors and forced through the recovery plant, which it leaves at a pressure of from 8 to 10 inches of water. The quantity of gas available depends entirely on the size of the ovens, and must be measured in each particular case.

It must also be borne in mind that on Sundays it is not usual to charge the ovens between 8 a.m. and 6 p.m., and that consequently there is a considerable falling off in the quantity of gas evolved during these hours.

Composition of the gas. Nitrogen is in excess of the amount which can be derived from the coal itself. This is explained by the fact that a proportion of air leaks into the ovens. The composition of the gas apart from this nitrogen is that of ordinary coal gas. The following table gives this composition taken over one day in each week over five weeks:

TABLE A.

Composition of gases.

	1st week	2nd week	3rd week	4th week	5th week
Vapours absorbed by alcohol		5·1*			
Carbon dioxide	4·53	1·9	2·88	4·1	5·13
Heavy hydrocarbons	2·20	0·1	1·60	1·35	1·8
Oxygen			0·91	3·31	1·99
Carbon monoxide	6·71	8·4	5·66	5·42	5·39
Hydrogen	33·62	31·79	47·07	27·46	29·17
Marsh gas	26·57	22·36	26·06	22·85	24·53
Nitrogen	26·74	30·40	15·68	34·79	32·4

* The alcohol apparently absorbs carbon dioxide and heavy hydrocarbons, consequently this absorption was not repeated.

Calorific power. The calorific power is also very variable, as would be expected from the variable composition of the gas. The higher and lower calorific values ascertained during tests over three different days are given in Table B.

In Table C the theoretical calorific power calculated from the analysis in Table A is given for the purpose of comparison.

Density of the gas. The density of the gas is 0·522 (air = 1). 100 c. ft. of the gas at 32° F. and 30″ pressure weigh 4·212 lbs.

These numbers will vary slightly with the composition but not to any very great extent.

Temperature of the gas. The gas usually leaves the recovery plant practically at atmospheric temperature.

Impurities. The quantity of sulphur and tarry matters in the gas is very considerable. The average amount of the former is as follows:

360 grains of sulphur for one hundred cubic feet of gas. On occasion, however, the amount of sulphur is as high as 600 grains per 100 c. ft. The amount of tarry matter measures from 2·69 to 3 grains per 100 c. ft.

Purification by effective tar-extractors, preferably mechanically driven, are necessary before the gas can be used for engine work.

TABLE B.

Calorific power of one cubic foot of gas in B.Th.U.

	1 cub. ft. at 32° F. & 30″		1 cub. ft. at 60° F. & 30″	
	Higher B.Th.U.	Lower B.Th.U.	Higher B.Th.U.	Lower B.Th.U.
1st day	422	386	407	370
	440	404	422	387
	422	386	407	370
	464	428	445	410
	440	404	422	387
	440	404	422	387
2nd day	403	367	386	351
	403	367	386	351
	403	367	386	351
3rd day	407	382	390	366
	396	372	379	356
	396	372	379	356
	396	360	379	345
	408	372	391	356
	407	372	390	355
Maximum	464	428	445	410
Minimum	396	360	379	345
Average	415	382	399	366

TABLE C.

	1st week, B.Th.U.	2nd week, B.Th.U.	3rd week, B.Th.U.	4th week, B.Th.U.	5th week, B.Th.U.
Theoretical calorific power calculated from analysis ...	448	439	472	366	403

APPENDIX B.

BLAST FURNACE GAS.

SINCE iron ore is smelted by mixing with coke and subjecting both to an air blast in a furnace, it follows that a large volume of gas similar in composition to that of producer gas, such as would be made if no steam were used, is given off at the mouth of the furnace. By applying a suitable collector and hopper valve to the latter it has now become customary to collect this gas, which formerly was wasted, and to use the same for power and heating purposes, this being carried out without interfering with the proper charging of the furnace. The quality of the gas naturally varies through a considerable range, and Table D below has been prepared from a large series of analyses and shews what may usually be expected under working conditions.

TABLE D.

Composition of blast furnace gas.

Carbonic dioxide	6·45	6·01	6·54	6·25	5·48
Carbonic monoxide	23·22	25·13	32·14	33·12	29·27
Hydrogen............	6·45	2·73	5·35	5·31	3·65
Nitrogen	63·88	66·13	55·97	55·32	61·60
	100·00	100·00	100·00	100·00	100·00
Combustible gas...	29·67 °/。	27·86 °/。	37·49 °/。	38·43 °/。	32·92 °/。
British Thermal Units per cubic foot of gas	100	96·01	129·31	132·46	113·69

INDEX.

Cambridge:
PRINTED BY JOHN CLAY, M.A.
AT THE UNIVERSITY PRESS.

Electrical Traction.

BY ERNEST WILSON, WHIT. SCH., M.I.E.E.,
Professor of Electrical Engineering in the Siemens Laboratory, King's College, London,
AND FRANCIS LYDALL, B.A., B.Sc.

NEW EDITION. REWRITTEN AND GREATLY ENLARGED.

Two volumes, sold separately. Demy 8vo., cloth.

Vol. I., with about 270 Illustrations and Index.

Vol. II., with about 170 Illustrations and Index.

15s. net each volume (post free 15s. 5d.).

"We are most decidedly of the opinion that both of these volumes will prove of great value to engineers, and that the last volume, in view of the present great interest in the question of single-phase traction, is of the utmost importance, for in it for the first time is published a great amount of data with reference to which, hitherto, the manufacturing companies concerned have observed great secrecy."—*The Times (Engineering Supplement)*.

A Text-Book of Electrical Engineering.

BY DR. ADOLF THOMÄLEN.

Translated by G. W. O. HOWE, M.SC., WHIT. SCH., A.M.I.E.E.,
Lecturer in Electrical Engineering at the Central Technical College, South Kensington.

With 454 Illustrations. Royal 8vo., cloth, 15s. net (post free 15s. 6d.).

This translation of the "Kurze Lehrbuch der Electrotechnik" is intended to fill the gap which appears to exist between the elementary text-books and the specialized works on various branches of electrical engineering.

Alternating Currents.

A Text-Book for Students of Engineering.

BY C. G. LAMB, M.A., B.Sc.,
Clare College, Cambridge; Associate Member of the Institution of Electrical Engineers; Associate of the City and Guilds of London Institute.

viii + 325 pages. With upwards of 230 Illustrations. Demy 8vo., cloth, 10s. 6d. net (post free 10s. 11d.).

Electric and Magnetic Circuits.

BY ELLIS H. CRAPPER, M.I.E.E.,
Head of the Electrical Engineering Department in the University College, Sheffield.

viii+380 pages. Demy 8vo., cloth, 10s. 6d. net (post free 10s. 11d.).

Applied Electricity.

A Text-Book of Electrical Engineering for "Second Year" Students.

BY J. PALEY YORKE,
Head of the Physics and Electrical Engineering Department at the London County Council School of Engineering and Navigation, Poplar.

xii+420 pages. Crown 8vo., cloth, 7s. 6d.

Hydraulics.

By F. C. LEA, B.Sc., A.M.Inst.C.E.,

Senior Whitworth Scholar, A.R.C.S.; Lecturer in Applied Mechanics and Engineering Design, City and Guilds of London Central Technical College, London.

With 367 Diagrams. Demy 8vo., 18s. net (post free 18s. 5d.).

This book is intended to supply the want felt by students and teachers alike for a text-book of Hydraulics to practically cover the syllabuses of London and other Universities, and of the Institution of Civil Engineers.

Hydraulics.

By RAYMOND BUSQUET,

Professeur à l'École Industrielle de Lyon.

AUTHORIZED ENGLISH EDITION.

Translated by A. H. PEAKE, M.A.,

Demonstrator in Mechanism and Applied Mechanics in the University of Cambridge.

viii+312 pages. With 49 Illustrations. Demy 8vo., cloth, 7s. 6d. net (post free 7s. 10d.).

The Balancing of Engines.

By W. E. DALBY, M.A., B.Sc., M.Inst.C.E., M.I.M.E.,

Professor of Engineering, City and Guilds of London Central Technical College.

SECOND EDITION, REVISED AND ENLARGED.

xii+283 pages. With upwards of 180 Illustrations.

Demy 8vo., cloth, 10s. 6d. net (post free 10s. 10d.).

Valves and Valve Gear Mechanisms.

By W. E. DALBY, M.A., B.Sc., M.Inst.C.E., M.I.M.E.,

Professor of Engineering, City and Guilds of London Central Technical College.

xviii+366 pages. With upwards of 200 Illustrations.

Royal 8vo., cloth, 21s. net (post free £1 1s. 5d.).

Power Gas Producers.

Their Design and Application.

By PHILIP W. ROBSON,

Of the National Gas Engine Co., Ltd.; sometime Vice-Principal of the Municipal School of Technology, Manchester.

Demy 8vo., cloth, 10s. 6d. net (post free 10s. 11d.).

The recent enormous increase in the use of gas power is largely due to the improvements in gas producers. This book, which is written by a well-known expert, goes thoroughly into the theory, design, and application of all kinds of plants, with chapters on working and general management.

The Strength and Elasticity of Structural Members.

BY R. J. WOODS, M.E., M.INST.C.E.,

Formerly Fellow and Assistant Professor of Engineering, Royal Indian Engineering College, Cooper's Hill.

SECOND EDITION, REVISED.

xii + 310 pages. With 292 Illustrations. Demy 8vo., cloth, 10s. 6d. net (post free 10s. 10d.).

Calculus for Engineers.

BY JOHN PERRY, M.E., D.SC., F.R.S.,

Professor of Mechanics and Mathematics in the Royal College of Science, London; Vice-President of the Physical Society; Vice-President of the Institution of Electrical Engineers.

NINTH IMPRESSION.

viii + 382 pages. With 106 Illustrations. Crown 8vo., cloth, 7s. 6d.

Mathematical Drawing.

Including the Graphic Solution of Equations.

BY G. M. MINCHIN, M.A., F.R.S.,

Formerly Professor of Applied Mathematics at the Royal Indian Engineering College, Cooper's Hill;

AND JOHN BORTHWICK DALE, M.A.,

Assistant Professor of Mathematics at King's College, London.

Crown 8vo., cloth, 7s. 6d. net (post free 7s. 10d.).

Five-Figure Tables of Mathematical Functions.

Comprising Tables of Logarithms, Powers of Numbers, Trigonometric, Elliptic, and other Transcendental Functions.

BY JOHN BORTHWICK DALE, M.A.,

Assistant Professor of Mathematics at King's College, London.

vi + 92 pages. Demy 8vo., cloth, 3s. 6d. net (post free 3s. 10d.).

This collection of Tables has been selected for use in the examinations of the University of London.

Logarithmic and Trigonometric Tables (To Five Places of Decimals). By JOHN BORTHWICK DALE, M.A., Assistant Professor of Mathematics at King's College, London. Demy 8vo., cloth, 2s. net (post free 2s. 3d.).

Traverse Tables. With an Introductory Chapter on Co-ordinate Surveying. By HENRY LOUIS, M.A., Professor of Mining and Lecturer on Surveying, Armstrong College, Newcastle-on-Tyne; and G. W. CAUNT, M.A., Lecturer in Mathematics, Armstrong College, Newcastle-on-Tyne. xxviii + 92 pages. Demy 8vo., flexible cloth, rounded corners, 4s. 6d net (post free 4s. 9d.).

Organic Chemistry for Advanced Students.

BY JULIUS B. COHEN, PH.D., B.SC.,

Professor of Organic Chemistry in the University of Leeds, and Associate of Owens College, Manchester.

Demy 8vo., cloth, 21s. net (post free 21s. 6d.).

The book is written for students who have already completed an elementary course of Organic Chemistry, and is intended largely to take the place of the advanced text-book. For it has long been the opinion of the author that, when the principles of classification and synthesis and the properties of fundamental groups have been acquired, the object of the teacher should be, not to multiply facts of a similar kind, but rather to present to the student a broad and general outline of the more important branches of the subject. This method of treatment, whilst it avoids the dictionary arrangement which the text-book requires, leaves the writer the free disposal of his materials, so that he can bring together related substances, irrespective of their nature, and deal thoroughly with important theoretical questions which are often inadequately treated in the text-book.

The Chemical Synthesis of Vital Products and the Inter-relations between Organic Compounds.

BY RAPHAEL MELDOLA, F.R.S., V.P.C.S., F.I.C., etc.,

Professor of Chemistry in the City and Guilds of London Technical College, Finsbury.

Vol. I., xvi+338 pages. Super Royal 8vo., cloth, 21s. net (post free 21s. 5d.).

The great achievements of modern Organic Chemistry in the domain of the synthesis or artificial production of compounds which are known to be formed as the result of the vital activities of plants and animals have not of late years been systematically recorded. The object of the present book is to set forth a statement, as complete as possible, of the existing state of knowledge in this most important branch of science.

The Chemistry of the Diazo-Compounds.

BY JOHN CANNELL CAIN, D.Sc. (Manchester and Tübingen),

Editor of the Publications of the Chemical Society.

176 pages. Demy 8vo., 10s. 6d. net (post free 10s. 10d.).

Lectures on Theoretical and Physical Chemistry.

BY DR. J. H. VAN 'T HOFF,

Professor of Chemistry at the University of Berlin.

Translated by R. A. LEHFELDT, D.SC.,

Professor of Physics at the Transvaal Technical Institute, Johannesburg.

In three volumes, demy 8vo., cloth, 28s. net (post free 28s. 7d.), or separately as follows:

PART I. CHEMICAL DYNAMICS. 254 pages, with 63 Illustrations. 12s. net (post free 12s. 5d.).

PART II. CHEMICAL STATICS. 156 pages, with 33 Illustrations. 8s. 6d. net (post free 8s. 10d.).

PART III. RELATIONS BETWEEN PROPERTIES AND COMPOSITION. 143 pages, 7s. 6d. net (post free 7s. 10d.).

Experimental Researches with the Electric Furnace.

BY HENRI MOISSAN,

Membre de l'Institut; Professor of Chemistry at the Sorbonne.

AUTHORIZED ENGLISH EDITION.

Translated by A. T. de MOUILPIED, M.Sc., Ph.D.,

Assistant Lecturer in Chemistry in the University of Liverpool.

xii + 307 pages, with Illustrations. Demy 8vo., cloth, 10s. 6d. net (post free 10s. 10d.).

"There is hardly a page of it which is not crowded with interest, and hardly a section which does not teem with suggestion; and if the coming of this English edition of the book has been so long delayed, we may still be thankful that it has come at last, and come in a form which it is a pleasure to handle and a delight to read."—*Electrical Review.*

Electrolytic Preparations.

Exercises for use in the Laboratory by Chemists and Electro-Chemists.

BY DR. KARL ELBS,

Professor of Organic and Physical Chemistry at the University of Giessen.

Translated by R. S. HUTTON, M.Sc.,

Demonstrator and Lecturer on Electro-Chemistry at the University of Manchester.

xii + 100 pages. Demy 8vo., cloth, 4s. 6d. net (post free 4s. 10d.).

The book contains a complete course of examples on the application of electrolysis to the preparation of both inorganic and organic substances. It will be found useful as filling a distinct gap in the text-book literature suitable for use in chemical laboratories, and should enable the chemist to make use of the many valuable and elegant methods of preparation which have been worked out during recent years, the advantages and ease of application of which he cannot appreciate without such a guide.

Introduction to Metallurgical Chemistry for Technical Students.

BY J. H. STANSBIE, B.SC. (LOND.), F.I.C.,

Associate of Mason University College, and Lecturer in the Birmingham University Technical School.

SECOND EDITION.

xii + 252 pages. Crown 8vo., cloth, 4s. 6d.

An Experimental Course of Chemistry for Agricultural Students. By T. S. DYMOND, F.I.C., Lately Principal Lecturer in the Agricultural Department, County Technical Laboratories, Chelmsford. New Impression. 192 pages, with 50 Illustrations. Crown 8vo., cloth, 2s. 6d.

A History of Chemistry.

BY DR. HUGO BAUER,

Royal Technical Institute, Stuttgart.

Translated by R. V. STANFORD, B.Sc. (LOND.),

Priestley Research Scholar in the University of Birmingham.

Crown 8vo., cloth, 3s. 6d. net.

The Becquerel Rays and the Properties of Radium.

BY THE HON. R. J. STRUTT, F.R.S.,
Fellow of Trinity College, Cambridge.

SECOND EDITION, REVISED AND ENLARGED.

viii + 222 pages, with Diagrams. Demy 8vo., cloth, 8s. 6d. net (post free 8s. 10d.).

"If only a few more books of this type were written, there might be some hope of a general appreciation of the methods, aims, and results of science, which would go far to promote its study. . . . A book for which no praise can be excessive."—*Athenæum.*

Astronomical Discovery.

BY HERBERT HALL TURNER, D.Sc., F.R.S.,
Savilian Professor of Astronomy in the University of Oxford.

xii + 225 pages, with Plates and Diagrams. Demy 8vo., cloth, 10s. 6d. net (post free 10s. 11d.).

An Introduction to the Theory of Optics.

BY ARTHUR SCHUSTER, Ph.D., Sc.D., F.R.S.,
Recently Professor of Physics at the University of Manchester.

xvi + 340 pages, with Illustrations. Demy 8vo., cloth, 15s. net (post free 15s. 5d.).

"We know of no book written with a set purpose better adapted to serve the purpose for which it was written, nor any that the earnest student of optics will find more interesting and profitable. The work itself, without the confession of the preface, shows that Professor Schuster is a teacher, and every page bears evidence that he is a master of his subject. . . . We heartily recommend the book to our readers."—*Ophthalmic Review.*

Wood.

A Manual of the Natural History and Industrial Applications of the Timbers of Commerce.

BY G. S. BOULGER, F.L.S., F.G.S., A.S.I.,
Professor of Botany and Lecturer on Forestry in the City of London College, and formerly in the Royal Agricultural College.

NEW EDITION. Revised and Enlarged and profusely illustrated. Demy 8vo., 12s. 6d. net (post free 12s. 11d.).

"It is just the book that has long been wanted by land agents, foresters, and woodmen, and it should find a place in all technical school libraries."—*Field.*

Manual of Alcoholic Fermentation and the Allied Industries.

BY CHARLES G. MATTHEWS, F.I.C., F.C.S., ETC.

xvi + 295 pages, with 8 Plates and 40 Illustrations. Crown 8vo., cloth, 7s. 6d. net (post free 7s. 10d.).

"This is a book worthy of its author, and well worth perusing by every student. . . . The student, both old and young, as well as the practical brewer, will find this book gives him some very useful information."—*Brewers' Guardian.*

The Evolution Theory. By DR. AUGUST WEISMANN, Professor of Zoology in the University of Freiburg in Breisgau. Translated, with the Author's co-operation, by J. ARTHUR THOMSON, Regius Professor of Natural History in the University of Aberdeen; and MARGARET THOMSON. Two vols., xvi + 416 and viii + 396 pages, with over 130 Illustrations. Royal 8vo., cloth, 32s. net.

"The subject has never been so fully and comprehensively expounded before; and it is not necessary to subscribe to all the author's tenets in order to recognise the value and the absorbing interest of his exposition, with its prodigious wealth of illustration, its vast store of zoological knowledge, its ingenious interpretations and far-reaching theories. English readers have reason to be grateful to Professor and Mrs. Thomson for their admirable translation."—*The Times.*

The Chances of Death and Other Studies in Evolution. By KARL PEARSON, M.A., F.R.S., Professor of Applied Mathematics in University College, London, and formerly Fellow of King's College, Cambridge. 2 vols., xii + 388 and 460 pages, with numerous Illustrations. Demy 8vo., cloth, 25s. net (post free 25s. 6d.).

The Life of the Salmon. With reference more especially to the Fish in Scotland. By W. L. CALDERWOOD, F.R.S.E., Inspector of Salmon Fisheries for Scotland. Illustrated. Demy 8vo., 7s. 6d. net.

"We have no hesitation whatever in advising all persons interested in the salmon, whether as fishermen, naturalists, or legislators, to add this book to their libraries." —*Nature.*

Animal Behaviour. By Professor C. LLOYD MORGAN, LL.D., F.R.S., Principal of University College, Bristol. viii + 344 pages, with 26 Illustrations. Crown 8vo., cloth, 7s. 6d. net (post free 7s. 11d.).

This important contribution to the fascinating subject of animal psychology covers the whole ground from the behaviour of cells up to that of the most highly developed animals.

BY THE SAME AUTHOR.

Habit and Instinct. viii + 352 pages, with Photogravure Frontispiece. Demy 8vo., cloth, 16s.

Professor ALFRED RUSSEL WALLACE:—"An admirable introduction to the study of a most important and fascinating branch of biology, now for the first time based upon a substantial foundation of carefully observed facts and logical induction from them."

BY THE SAME AUTHOR.

The Springs of Conduct. Cheaper Edition. viii + 317 pages. Large crown 8vo., cloth, 3s. 6d. This volume deals with the Source and Limits of Knowledge, the Study of Nature, the Evolution of Scientific Knowledge, Body, and Mind, Choice, Feeling, and Conduct.

BY THE SAME AUTHOR.

Psychology for Teachers. New Edition, entirely rewritten. xii + 308 pages. Crown 8vo., cloth, 4s. 6d.

An Introduction to Child Study. By Dr. W. B. DRUMMOND. Crown 8vo., cloth, 6s. net.

The Child's Mind: Its Growth and Training. By W. E. URWICK, University of Leeds. Crown 8vo., cloth, 4s. 6d. net.

2848152

Made in the USA